ELECTRONIC SCANNING RADAR SYSTEMS (ESRS) DESIGN HANDBOOK

Peter J. Kahrilas
Missile Systems Division
Raytheon Company

ELECTRONIC SCANNING RADAR SYSTEMS (ESRS) DESIGN HANDBOOK

ARTECH HOUSE

Printed and bound in the United States of America.

Standard Book Number: 0-89006-023-1

Library of Congress Catalog Card Number: 75-43051

This book is dedicated to

Leticia
Aldo R. Miccioli
David K. Barton

who provided the encouragement and support to accomplish this work.

Dedication

Foreword

The literature on radar contains many books and articles on systems engineering, and more recently a large amount of material on phased array antennas and electronic scanning techniques. What has been lacking until now is a practical guide to design of radar systems using electronically scanned beams, and it was to fill this gap that the present volume was written.

The original radars of the 1930's used array antennas, but their phase distributions were fixed and the beams were scanned mechanically. Fast electro-mechanical scanners were developed for microwave radar during World War II, in which the phase shift of individual radiating elements was varied to scan the beam continuously at rates up to hundreds of degrees per second. During the early 1950's, the technique of frequency scanning was developed and used in practical search radars. In the late 1950's, designs for two-dimensional phased arrays were developed which made possible electronic scan over a broad volume in space, with hemispherical coverage provided by three or four individual antennas.

Array radar systems did not appear automatically in the wake of array antenna designs. The high cost of the antenna elements and associated feed and control systems limited early applications to the most critical and demanding roles, such as ballistic missile defense and air defense of high-value ships. The multifunction radar approach to meet simultaneously (or in rapid, time-shared fashion) the diverse requirements of search, tracking, and target engagement in a multiple-target environment; and of a scheduling and control system to coordinate the radar system has proven to be a formidable task.

It is this system engineering task for the electronic scanning radar which is the subject of this book. Because of the need to balance the allocation of radar resources between functions, and to compromise between functions in selection of major radar parameters, the task of the electronically scanned radar system designer is particularly demanding. The techniques and procedures outlined here will be of great assistance in arriving at the best design for the given problem, in the most direct way and with an appreciation of the trade-offs involved in matching the radar to its requirements.

David K. Barton

Preface

This book is primarily intended for that rare species of man known as the Electronic Scanning Radar System (ESRS) "Architect." It also provides an overview of the wonderful world of ESRS. Electronic scanning radar system design challenges the boundaries of radar technology and makes maximum use of the radar range equation and radar performance analysis — when performed correctly.

Although this book is primarily intended for ESR system designers or architects, radar subsystem and specialty groups all have a vested interest. The array designer is interested in relating array design to systems; single beam versus multiple beam, phase or frequency scan, optical or corporate feed, phase shifters (diode or ferrite) or real time devices. The transmitter designer is concerned whether he should develop high power single tube transmitters, multiple intermediate size transmitter tubes, or solid state transmit/receive modules; and so forth. An electronic scanning radar system design book provides these specialty groups with perspective on where they can best apply their talents for present and future designs.

The ESRS Architect evolves the final "balanced" system design by a systematic procedure that draws on the contributions of performance analysts, waveform experts, array, beam steering unit, transmitter, receiver, signal processor, computer and peripherals, radar interface unit, control console/displays, reliability, maintainability, and quality control specialists; to meet all the performance requirements at minimum: risk, complexity, cost and development schedule. The approach is not unique nor are the results indisputably optimum.

This is the only book to my knowledge that deals directly with the development of a complete electronic scanning radar system design. There are many excellent texts on radar analysis, components, techniques and systems; but none that develop a system design methodology leading to the evolution of an electronic scanning radar system to operate in a particular environment and satisfy the requirements specified by the potential user. The book evolved from papers and lectures given by the author. The final impetus to publish came in the spring of 1974 when the author delivered a two week series of lectures in an intensive course on electronic scanning radar system design at the Indian Institute of Technology — Kanpur. Several of the participants, representatives of India's academic, military and industrial communities; urged the author to write on system applications of arrays.

The book is structured to provide in a logical progression the tools required by a system architect to develop an integrated/ balanced electronic scan radar system design. Although there is in every chapter some discussion on characteristic(s) peculiar to electronic scanning radar system design, most of the concepts apply to the development of conventional radar system design as well. In Chapter 1 the radar range equation is derived emphasizing the underlying assumptions and it is reformatted for convenient use in ESRS design. Types and typical values of ESR range equation losses are given and the chapter is concluded with an illustrative example. Chapter 2 indicates how electronic scanning radar systems are particularly suited to meet modern radar requirements and provides a basis for selection between conventional and electronic scanning radars. The third chapter describes electronic scanning radar techniques, limitations, and applications. Chapters 4 through 6 deal with detection, matched filter, waveform, and Moving Target Indication (MTI) theory with emphasis on electronic scanning radar application. The many uses of sine space coordinates in electronic scanning radar system design are discussed in Chapter 7, including an illustrative example. Tracking considerations unique to electronic scanning radar system; such as: sampled data servo loops in all three coordinates with low sampling rates on multiple targets on a time-shared basis and interpolative-null tracking, are discussed in Chapter 8 along with recursive formulas for Type II and Type III filters. Chapter 9 covers accuracy and resolution design criteria with use of millisines instead of milliradians in angular coordinates for electronic scanning radar applications. The estimation expressions for the basic error sources in electronic scanning radars are conveniently summarized in tabular form for use by the system performance analyst. Parameter management, described in Chapter 10, is a method the author developed some 15 years ago to obtain the maximum performance for minimum power-aperture and design complexity in an electronic scanning radar system by judicious control of all radar parameters. Chapter 11 contains the major electronic scanning radar parameters for designing and defining a system as well as operational scenario, time, power, error, and loss budgets. By Chapter 12 the reader is prepared to appreciate and use a system design methodology. Chapter 13 deals with ESRS integration, test and evaluation, with emphasis on the procedures and controls required for successful accomplishment. The concluding Chapter 14 contains a case study that proceeds step by step through a complete electronic scanning radar system design, illustrating the use of the material discussed in the preceding chapters. The case study begins with the performance requirements for a hypothetical threat model and concludes with a detailed listing of electronic scanning radar system parameters evolving from the design that has been developed to satisfy the requirements.

Finally, there is the belief by the author, many of his peers, and the audiences addressed; that there is a significant dearth of information in the literature on the subject of electronic scanning radar system design. This book is a sincere attempt to partially fill this void.

Peter J. Kahrilas
April 1975

Acknowledgements

The author is grateful to the Raytheon Company for creating the environment in which this work could be performed and granting permission to publish this manuscript. In particular, D. S. Banks, Chief Scientist; Dr. P.W. Cheney, Manager of the Bedford Laboratories; and W. W. McLeod, Chief Engineer.

My additional indebtedness to David K. Barton for his useful suggestions throughout the book during his review of the manuscript. At various times I have benefited from discussions with members of the Senior Technical Staff and Radar Laboratory Staff. Dr. I. Kanter has been the source of clarification for numerous mathematical concepts. A special note of thanks is due to Demetra Flanagan, Louise Gryszkiewicz, Mary Wayne, and Donna Larson for typing the manuscript and Andrew Reynolds for the photographic artwork preparation.

The author is equally indebted to the many associates in government agencies, universities and industry for permission to reproduce photographs of their electronic scanning radars. Thanks are also due to the authors of technical books and articles where the reader is directed for more rigorous treatment of some of the referenced material.

Table of

Contents

Chapter One

Chapter Two

Chapter Three

Chapter Four

Chapter Five

Chapter Six

Chapter Seven

Chapter Eight

Chapter Nine

Chapter Ten

Chapter Eleven

Chapter Twelve

Chapter Thirteen

Chapter Fourteen

Fundamental Radar Concepts

Chapter One

1.1 Primary Radar Functions — Radar (RAdio Detection And Ranging), as the palindrome acronym implies, is an electronic device for the detection, location (and identification) of objects. Radar:

"Sees" (detects) objects in the presence of irrelevant environment such as ground clutter, rain, haze, fog, snow, clouds, smoke, and in darkness. It can also be designed to see through (or detect) these conditions impervious to normal human vision. It is not a substitute for the human eye, since the eye can resolve detail better than can radar and can recognize colors, but it cannot see through irrelevant environment nor measure. However, it extends the capability of man's senses for observing his environment.

Measures range (the most important radar function) and angles of detected objects. Doppler velocity, f_D, is measurable as a fourth basic quantity. Targets can be located and their trajectories computed from these coordinates (R, A_z, E) and their time derivatives ($\dot{R}$ or f_D, $\dot{A}_z$, $\dot{E}$) and ($\ddot{R}$ or $\dot{f}_D$, $\ddot{A}_z$, $\ddot{E}$).

Measures amplitude of target echo(es) that relate to object size and shape in combination with the other radar measurements which provide target identification.

1.2 Description of a Pulse Radar System — A simplified block diagram of a pulse radar system is shown in Fig. 1.1.

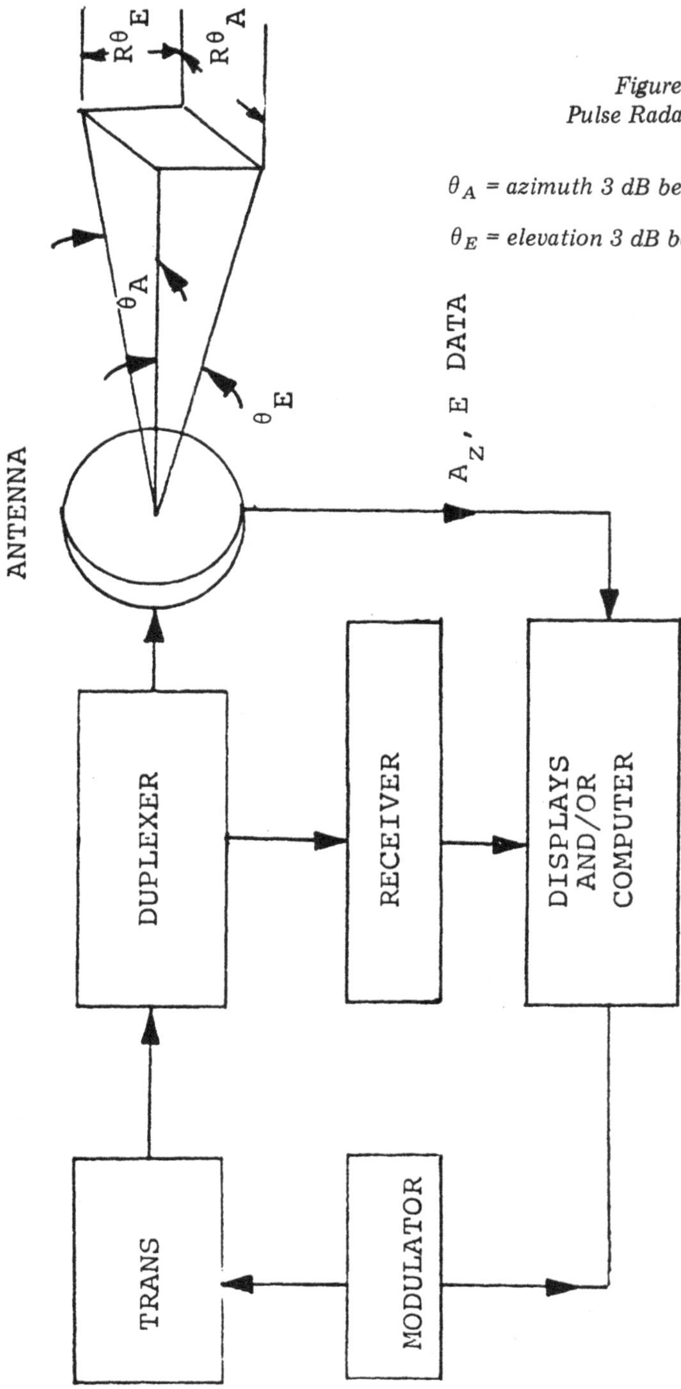

Figure 1-1
Pulse Radar System

θ_A = *azimuth 3 dB beamwidth*

θ_E = *elevation 3 dB beamwidth*

Time measurement provides range of target from radar. Since electromagnetic energy travels at the speed of light the range

$$R = \frac{ct}{2} \tag{1.1}$$

where the velocity of light $c = 3 \times 10^8$ m/s and t is the time taken by the pulse to travel to the target and return.

Therefore, measurement of range in a pulse radar system is made as a time delay between the transmitted signal and the received target echo signal. The sharp leading and trailing edges of the wide bandwidth pulse wave form serve as ideal references for time delay measurements.

TYPICAL VALUES

t	R
1 μs	150m = 164 yds = 492 ft
10.74 μs	1.0 statute mile
12.36 μs	1.0 nautical mile = 6076 ft

Speed is obtained by smoothing successive range rate measurements of the form:

$$\dot{R}_{12} = \frac{R_2 - R_1}{t_o} \tag{1.2}$$

or by doppler frequency measurement,

$$f_D = \frac{2V}{\lambda} \tag{1.3}$$

where:

R_1, R_2 = two successive range measurements
t_o = interpulse period
V = target radial velocity
λ = wavelength of transmitted energy

Acceleration is obtained by:

$$\ddot{R} = \frac{\dot{R}_{23} - \dot{R}_{12}}{t_o} \quad \text{or} \quad \dot{f}_D = \frac{f_{D2} - f_{D1}}{t_o} \tag{1.4, 1.5}$$

(1.4)
(1.5)

The radar operation is initiated by the modulator pulsing the transmitter. Intense, brief, RF-electromagnetic wave bursts are generated at relatively long intervals of time. The pulsed RF energy from the transmitter passes through the duplexer to the antenna and is propagated into space. When a target exists in the direction of propagation, an echo signal is reflected back to the antenna. The target echo signal is collected by the antenna and passed through the duplexer to the receiver. After amplification and video detection, the signal is displayed or processed in the computer. The measurement of time between transmit and receive signals is performed for range determination either on a calibrated display, by the computer clock, or by a precision reference clock.

The direction of the target relative to the radar is determined from the direction of the radiated beam which in turn is usually obtained from the azimuth, A_z, and elevation, E, angular coordinates of the antenna. Alternately, the target direction and position can be obtained by trilateration using range measurements of two radars. The computer (if used) provides automatic control of the radar and trajectory computation.

1.3 Range Equation Derivation — The range equation derivation is repeated here briefly because of its fundamental importance to radar system design and to emphasize the underlying assumptions which affect the accuracy of range prediction.

Assume that the antenna shown in Fig. 1.1 has a rectangular beam shape

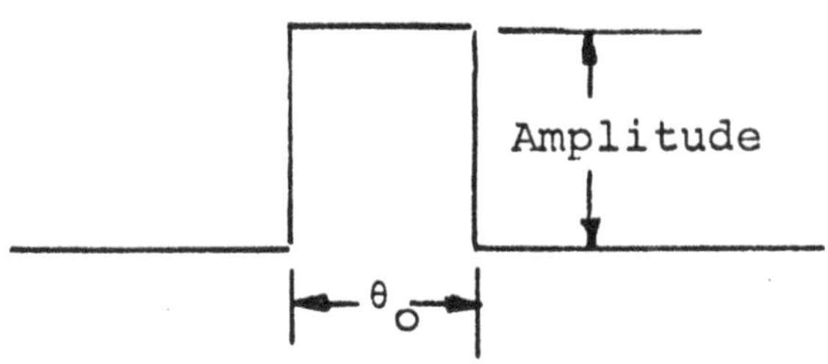

where the antenna beamwidth

$$\theta_o \approx \frac{\lambda}{D} \tag{1.6}$$

(Actually $\theta_o = \frac{.88\lambda}{D}$ to $\frac{1.3\lambda}{D}$ depending on the illumination taper)

D = aperture dimension

or

$$D \approx \frac{\lambda}{\theta_o}$$

Antenna area

$$A_o = D_E \, D_A \tag{1.7}$$

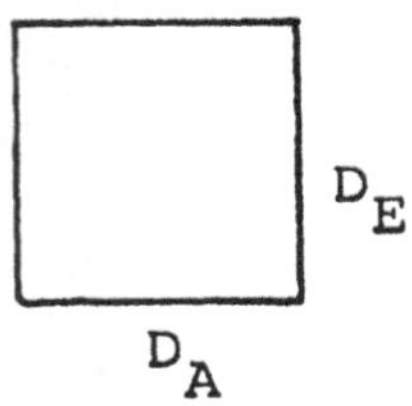

or

$$A_o \approx \frac{\lambda^2}{\theta_E \, \theta_A} \tag{1.8}$$

At a range R, an isotropic radiator (omnidirectional antenna) distributes its transmitted power with an equal intensity in all directions of space and the radiated power passes through the full surface area, $4\pi R^2$, of the imaginary sphere surrounding the isotropic radiator.

The gain from using the directive antenna is equal to the ratio of the surface area of the sphere to the area within the beam:

$$G = \frac{4\pi R^2}{R\theta_E \, R\theta_A} \tag{1.9}$$

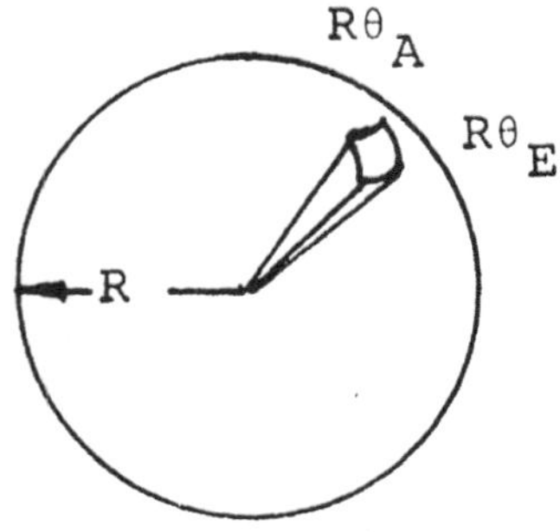

Substituting expression (1.8)

Antenna Area Gain

$$G = \frac{4\pi A_o}{\lambda^2} \qquad (1.10)$$

(For comparison, the directive antenna gain is

$$G = \frac{4\pi\ A_e}{\lambda^2}$$ [Ref. 1, Page 54]

where:

A_e = maximum effective aperture.)

If the peak power, P_p, of a radar transmitter is radiated through an isotropic antenna:

$$\text{Power density from isotropic radiator at range R} = \frac{P_p}{4\pi R^2} \qquad (1.11)$$

If instead a directive antenna of gain, G_T, is used to channel most of the radiated power, P_p, in the direction of the target:

$$\text{Power density from directive antenna at range R} = \frac{P_p\ G_T}{4\pi R^2} \qquad (1.12)$$

The target at range R intercepts a portion of the radiated power and reradiates (reflects) it in the direction of the radar. The portion of the power reflected by the target in the direction of the receiving antenna will depend on the radar target cross section, σ, otherwise known as the effective echoing area. Definition: If all the power incident on this effective echoing area were scattered uniformly in all directions with the same polarization as that of the receiving antenna, the received echo would then be equal to the actual echo from the target.

$$\therefore \text{Power intercepted by target} = \frac{P_p}{4\pi R^2} G_T \sigma = \text{reflected power} \qquad (1.13)$$

and

Peak power echo intercepted by receive antenna

$$S = \frac{P_p G_T \sigma}{4\pi R^2} \times \frac{1}{4\pi R^2} \times A_R = \frac{P_p G_T \sigma A_R}{(4\pi)^2 R^4} \qquad (1.14)$$

↑
isotropic reflection from target

but

$$A_R = \frac{G_R \lambda^2}{4\pi}$$ from expression (1.10)

$$\therefore S = \frac{P_P G_T G_R \sigma \lambda^2}{(4\pi)^3 R^4} \qquad (1.15)$$

$$S = \frac{P_P G^2 \sigma \lambda^2}{(4\pi)^3 R^4} \qquad (1.16)$$

If same antenna is used to transmit and receive $G_T = G_R = G$

Expressions (1.14), (1.15) and (1.16) are exact and do not depend on square aperture assumptions. Expression (1.10) is exact if A_R is defined as the effective area of the receive antenna.

Noise — The ability of a radar to detect a target echo signal is limited by the noise energy that occupies the same frequency spectrum as the signal energy. In microwave region (200 - 60000 MHz) man-made and antenna noise from space are secondary. Thermal noise is the main factor restricting radar performance capability.

$$\text{Noise Power} = N = kT_sB_N \qquad (1.17)$$

where:

k = Boltzman's constant, 1.38×10^{-23} watts/(Hz x degree Kelvin)

T_s = system noise temperature (degree Kelvin) referred to antenna face

B_N = noise bandwidth of the IF filter ($1/\tau$ is used for calculations, based on the matched filter, where τ = pulse width)

System noise temperature referred to antenna face

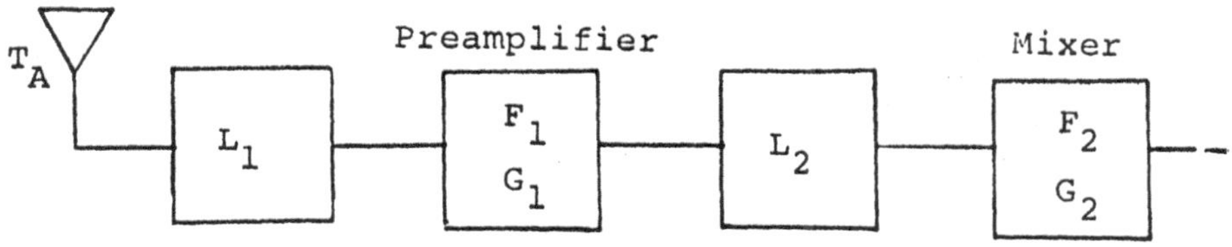

$$T_s = T_A + T_o \left[(L_1 - 1) + L_1 (F_1 - 1) + \frac{L_1 (L_2 - 1)}{G_1} + \frac{L_1 L_2 (F_2 - 1)}{G_1} + \ldots \right] \quad (1.18)$$

usually $G_1 \geqslant 20$ dB

and

$$T_s \approx T_A + T_o (L_1 F_1 - 1) \quad (1.19)$$

where

T_A = antenna sky noise temperature — degrees Kelvin (galaxy plus oxygen and water vapor)

T_o = ambient temperature — degrees Kelvin (usually 290°K)

L_1 = ohmic receive antenna plus microwave losses (antenna — radome, feed . . .) (microwave — rotary joint, duplexer/circulator, waveguide/cable, RF switch, TR switch and/or limiter)

F_1 = preamplifier noise figure

G_1 = preamplifier gain

L_2 = losses between preamplifier and mixer (bandpass filter, RF switch, waveguide/cable)

F_2 = mixer noise figure

G_2 = mixer conversion gain/loss

The radar equation for a practical system is obtained by dividing expression (1.16) by expression (1.17) and including a loss factor, L, to account for system losses.

$$\frac{S}{N} = \frac{P_p G_T \; G_R \; \sigma \, \lambda^2}{(4\pi)^3 R^4 \; LkT_s B_N} \tag{1.20}$$

The confusion associated with the various definitions of antenna gain and antenna efficiency is eliminated by the use of the following set of definitions for the terms in the radar equation:

G_T = transmit antenna area gain = $\dfrac{4\pi \; A_T}{\lambda^2}$

G_R = receive antenna area gain = $\dfrac{4\pi \; A_R}{\lambda^2}$

A_T = physical area of transmit aperture — meters^2

A_R = physical area of receive aperture — meters^2

L = system losses not included in system noise temperature such as:

1. Transmit antenna losses — illumination taper, radome, spillover, blockage, feed
2. Receive antenna non-dissipative losses — illumination taper, spillover, blockage
3. System degradation
4. Propagation
5. Beam shape
6, Beam scanning
7. Microwave transmit losses
8. Mismatch filter loss
9. Signal processing losses

R = detection range — meters

S/N = r.f. signal-to-noise ratio of single pulse required to obtain detection

σ = radar target cross section — meters^2

λ = transmission wavelength — meters

P_p = transmitter peak power — watts

B_N = noise bandwidth of the i.f. filter — hertz

T_s = system noise temperature referred to antenna face — degrees Kelvin

k = Boltzman's constant = 1.38×10^{-23} watts/(Hz x degree Kelvin)

Detection — Because of random noise sources and fluctuation characteristics of target cross section, target detection is a statistical decision process.

In each space cell the radar can make two kinds of errors:

1. It can indicate a target is present when no target exists — this error occurs with relative frequency, P_N, where P_N is the probability of false alarm.
2. It can indicate no target is present when a target does exist — this error occurs with relative frequency, $(1 - P_D)$, where P_D is the probability of detection.

S/N is selected to satisfy the P_N and P_D requirements for particular target scintillation characteristics (target models are discussed in Chapter 4 — Detection Theory).

Fig. 1.2 illustrates the relationship of threshold level to the probability density functions for noise alone and signal plus noise. The false alarm rate decreases as the threshold level is increased (moved to the right) but the probability of detection also decreases.

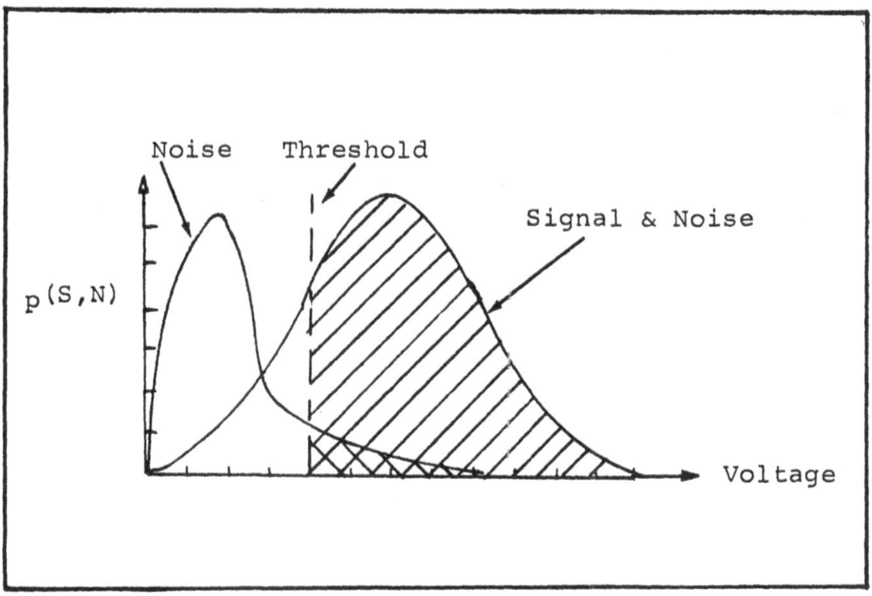

Figure 1-2
Probability Density Functions
For Noise Alone and Signal Plus Noise

Assumptions

1. The various fluctuation models (refer to Chapter 4 for listing and definitions) are realistic representations of radar targets and effectively relate false alarm probability and probability of detection to the r.f. single pulse signal-to-noise ratio used in the range equation to provide accurate range performance prediction.
2. The radar cross section, σ, is an adequate representation of complex targets (aircraft, missiles, and ships which are complicated functions of the viewing aspect and the radar frequency) to provide accurate range performance prediction.
3. The use of separate loss components in the system loss factor, L, (instead of evaluating an overall transmit/receive transfer function) adequately compensates the range equation parameters to provide accurate range performance prediction.
4. Departure from free space propagation is accounted for by system loss factor components (propagation, ground reflection lobing, beam shaping, scanning) and antenna gain modification (radome, illumination taper, spillover, blockage, mismatch, . . .).

1.4 **Range Equation Losses** — The most significant range equation losses and their typical values are itemized as a guide to the radar syster architect. The loss values given for radar components are quite representative of the current state-of-the-art. More detailed ESR loss contributors are indicated for specific designs in the various examples given throughout the text.

1. System degradation	1.0 — 6.0 dB
2. Antenna (illumination taper, radome, blockage, spillover)	1.0 — 4.5
3. Rotary Joint[1]	0.1 — 0.2
4. Duplexer[1] /Circulator	0.3 — 0.5
5. Harmonic Filter[1]	0.1 — 0.3
6. Feed[1] (monopulse or simple)	0.1 — 0.3
7. Waveguide[2]	0.1 —
8. Cable[3]	0.1 —
9. Isolator[1]	0.3 — 0.5

10. Hybrids	0.05 — 0.1
11. Band Pass Filter[1]	1.0 — 1.5
12. Mismatched Filter[4]	0.5 — 2.0
13. Signal Processing (CFAR, Limiting)	1.0 — 1.8
14. Beam Shape — fan beam — 1 dimension	1.6 (Ref. 2)
— pencil beam — 2 dimensions	3.2
15. Electronic Scanning[6]	$\sec^3 \xi$
16. Oxygen Absorption & Water Vapor[5] — 2 way / 0-10 GHz & 0-350 n mi (or more)	0 — 6.8 (Ref. 3)

Track —

1. Track Gate	1.0 — 1.6
2. Beam Shape (tracker stays near center of beam)	0.5

[1] function of frequency
[2] function of length and frequency
[3] function of length, width, dielectric material and frequency
[4] function of weighting/residue levels
[5] function of range through atmosphere, elevation angle, and frequency
[6] ξ is scan angle from array normal

System Noise Temperature (referred to antenna face)

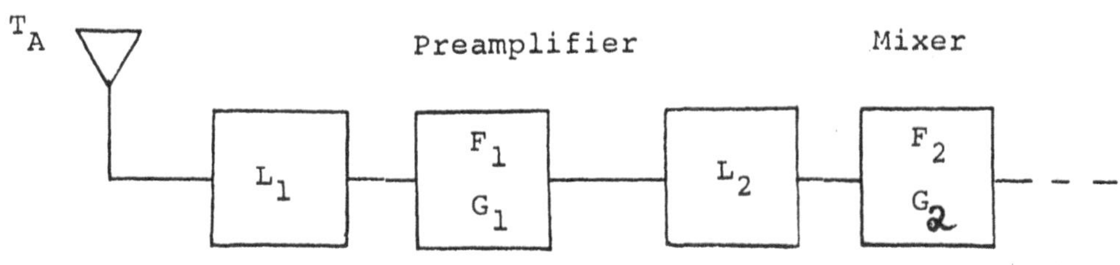

$$T_s = T_A + T_o \left[(L_1 - 1) + L_1 (F_1 - 1) + \frac{L_1 (L_2 - 1)}{G_1} + \frac{L_1 L_2 (F_2 - 1)}{G_1} + \ldots \right] \tag{1.18}$$

$$T_s \approx T_A + T_o (L_1 F_1 - 1) \qquad 250 - 3500^\circ K \tag{1.19}$$

T_A	= Sky noise temperature (Ref. 3) (atmospheric absorption plus galactic) (at 0° elevation angle)	(1 — 10 GHz) 90° — 160°K
L_1	— 1. Antenna (radome, feed [monopulse or simple]) — ohmic	0.5 — 2.8 dB
	2. Rotary Joint	0.1 — 0.2
	3. Duplexer/Circulator	0.3 — 0.5
	4. Waveguide/Cable	0.1 —
	5. T.R. Switch/Limiter	0.5 — 1.5
F_1	— Preamplifier Noise Figure	0.7 — 7.0
G_1	— Preamplifier Gain	15.0 — 30.0
L_2	— 1. Waveguide/Cable	0.1 —
	2. Bandpass Filter	1.0 — 1.5
F_2	— Mixer Noise Figure	6.0 — 8.0
G_2	— Mixer Conversion Loss/Gain	-6.0

1.5 **Energy (ESR) Form of Range Equation** — The energy form of the range equation is especially useful in the design of electronic scanning radar systems where single pulse detection is most commonly used; and energy and time are independently manipulated as functions of range, altitude, scanning angle, and operational mode. (Energy and time management are discussed in Chapter 10 — Parameter Management.)

Since

$$\text{Energy} \quad \mathcal{E} = P_p \, \tau \tag{1.21}$$

$$\text{or} \quad \mathcal{E} = P_A \, t_o \tag{1.22}$$

and

$$\text{Duty Cycle} = \frac{P_A}{P_p} = \frac{\tau}{t_o} = \tau f_r = \frac{f_r}{B_N} \tag{1.23}$$

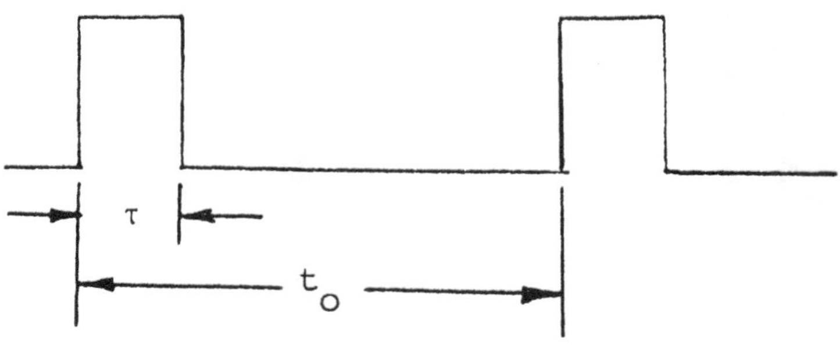

$$P_P = \frac{(4\pi)^3 \; R^4 \; S/N \; L \; kT_s}{\tau G_T \, G_R \, \sigma\lambda^2} \tag{1.24}$$

and

$$P_A = \frac{(4\pi)^3 \; R^4 \; S/N \; L \; k \; T_s}{t_o \, G_T \, G_R \, \sigma\lambda^2} \tag{1.25}$$

where:

$\mathcal{E}$ = transmitter energy per pulse — joules
P_P = transmitter peak power — watts
P_A = transmitter average power — watts
τ = pulse width — seconds
t_o = interpulse period — seconds
f_r = pulse repetition frequency — hertz

The peak and average power requirements are determined for individually tailored ESR designs. Power and time budgets are developed and balanced for all operational modes — search, multitarget acquisition, multitarget track, and multimissile illumination and guidance. For example, expressions (1.24) and (1.25) for the range equation are not readily applicable to ESR design because the pulse repetition frequency, noise bandwidth/pulse width, and angular scan rate are not necessarily constant throughout the search volume. To determine the average power required for an ESR search mode where one pulse is transmitted per beam position and n beam positions are required to cover the search volume:

$$P_{A_s} = \frac{\mathcal{E}_s}{t_s} = \frac{\sum_{i=1}^{n} \mathcal{E}_i}{\sum_{i=1}^{n} t_{o_i}} = \frac{\sum_{i=1}^{n} P_{p_i} \tau_i}{\sum_{i=1}^{n} t_{o_i}} \tag{1.26}$$

where:

P_{A_s} = average power in search mode — watts
$\mathcal{E}_s$ = total energy in search mode — joules
t_s = total frame time in search mode — seconds

and the subscript i denotes the value of that previously defined parameter for the ith beam position.

also

$$\text{Long term duty cycle} = \frac{\sum_{i=1}^{n} \tau_i}{\sum_{i=1}^{n} t_{o_i}} \tag{1.27}$$

and

$$\text{Short term duty cycle} = \frac{\tau_i}{t_{o_i}} \tag{1.28}$$

For added convenience the energy form of the range equation is derived for single-pulse detection containing a conversion constant, ν, to make it applicable with the mixed system of units most often employed in radar specifications.

$$\mathcal{E} = P_p \tau = \frac{(4\pi)^3 \; R^4 \; S/N \; L \; kT_s}{G_T \, G_R \, \sigma \lambda^2} \tag{1.29}$$

$$\mathcal{E} = \frac{R^4 \; S/N \; L(NF)_o}{\nu G_T \, G_R \, \sigma \lambda^2} \tag{1.30}$$

where

$$(NF)_o = \frac{\text{system noise temperature}}{\text{reference temperature}} = \frac{T_s}{T_o} = \frac{T_s}{290} = \begin{matrix}\text{operating}\\ \text{noise factor}\end{matrix} \tag{1.31}$$

and

$$\nu = \frac{1}{(4\pi)^3 \; k \, T_o} \times \left(\frac{\lambda^2 \sigma}{R^4} \; \begin{matrix}\text{conversion}\\ \text{units}\end{matrix} \right) \tag{1.32}$$

$$\nu = \frac{\lambda^2 (m^2/cm^2)}{(4\pi)^3 \; kT_o R^4 (m/nm)^4} \quad \text{for} \quad \begin{matrix}\lambda \text{ in cm}\\ R \text{ in nm}\\ \sigma \text{ in } m^2\end{matrix} \tag{1.33}$$

$$\nu = \frac{10^{-4}}{(4x3.14159)^3 x1.38047x10^{-23} x290x(1.85325)^4 x10^{12}}$$

$$\nu = \frac{10^{-4}}{1.9844x10^3 x1.38047x10^{-23} x2.9x10^2 x1.1796x10^{13}}$$

$$\nu = \frac{10^{-4}}{9.371075x10^{-5}} = 1.06711 = 0.28 \text{ dB} \qquad (1.34)$$

Units for Energy Equation (1.30) —

P_P	= watts	dBw
τ	= seconds	dBs
&	= joules	dBj
ν	= 1.06711	0.28 dB
R	= nautical miles	dBnm
$L, G_T, G_R, (NF)_o, S/N$	= ratio	dB
σ	= meters^2	dBm^2
λ^2	= centimeters^2	dBcm^2

1.6 Use of Range Equation —

Example: Determine the peak and average power required to detect aircraft at a range of 100 n mi with a single-pulse 90% probability of detection and no more than one false alarm every 100 hours. The aircraft fading characteristics are approximated by a Swerling I fluctuation model (defined in Chapter 4, Section 4.3) and the average radar cross section is 1.0 m^2.

1. *Determine Probability of False Alarm* — The probability of false alarm, P_N, is the probability that a single sample of noise will exceed the detection threshold of the radar receiver. If the receiver operates continuously with bandwidth, B_N, there will be B_N independent samples of noise per second applied to the threshold. Therefore, on the average:

$$\text{False alarms per second} = B_N P_N \tag{1.35}$$

or

$$\frac{1}{t_{fa}} = \underset{\substack{\text{range sweeps}\\\text{per second}}}{f_r} \times \underset{\substack{\text{number of}\\\text{independent}\\\text{samples per}\\\text{range sweep}}}{\frac{K_R\, t_o}{\tau}} \times P_N \tag{1.36}$$

$$\therefore P_N = \frac{\tau}{K_R\; t_{fa}} = \frac{1}{K_R\; B_N\; t_{fa}} \tag{1.37}$$

where:

t_{fa} = false alarm time

K_R = fraction of the range sweep that the receiver is gated on ($K_R \approx 1$)

Let $\tau = 36\ \mu s = 3.6 \times 10^{-5}\ s = -44.4\ \text{dBs}$

(Adjacent target resolution is approximately 6 n. miles. Targets are resolvable between 1.5 and 2.0 times the pulse width depending on the r.f. phase relationship of the target echo signals, steepness of the waveform skirts, and the sidelobe response.)

$$\therefore \quad P_N = \frac{3.6 \times 10^{-5}}{100\ \text{hrs} \times 3600\text{s/hr}} = 10^{-10}\ (K_R \approx 1)$$

2. *Determine single-pulse signal-to-noise ratio* for $P_N = 10^{-10}$ & $P_D = 0.90$ from Swerling I graphs or tables (Refs. 4 and 5).

   S/N = 23.4 dB

3. *Determine Area Gain* — Let beamwidth $\theta_o = 1^\circ$ (pencil beam) and operating frequency f_o = 3 GHz.

$$\theta_o = \frac{67.4\lambda}{D}$$

(Taylor taper, $\overline{n}$ = 4, 30 dB maximum sidelobe) (Ref. 6, Page 256) (Circular Array)

$$\lambda = \frac{c}{f_o} = \frac{3 \text{ x } 10^{10} \text{ cm/s}}{3 \text{ x } 10^9} = 10 \text{ cm}$$

$$\lambda^2 = 100 \text{ cm}^2 = 20.0 \text{ dB cm}^2$$

$$D = \frac{67.4\lambda}{\theta_o} = 67.4 \text{ x } 10 = 674 \text{ cm} = 6.74 \text{ m} = 22.1' \text{ aperture}$$

$$G = \frac{4\pi A}{\lambda^2} = \frac{\pi^2 D^2}{\lambda^2} = \frac{9.87 \text{ x } 6.74^2 \text{ x } 10^4}{10^2} = 4.48 \text{ x } 10^4 \text{ area gain}$$

$$G = 46.5 \text{ dB}$$

$$G_R G_T = G^2 = 93.0 \text{ dB}$$

4. Determine Operating Noise Factor —

$T_s = T_A + T_o(L_1 F_1 - 1)$

$T_o = 290° \text{ K}$

$T_A = 110° \text{ K}$ ($f_o = 3$ GHz, $E = 0°$ Ref. 3, Page 68)

$F_1 = 2.0$ dB (uncooled parametric amplifier) = 1.58

Given:

			Ohmic	Non-Ohmic
L_1 =	1. Antenna			
	Radome	=	0.5	
	Ill. Taper	=		1.6
	Spillover	=		1.2
	Feed	=	0.2 0.7	2.8
		0.7		
	2. Rotary Joint	0.1		
	3. Duplexer	0.5		
	4. TR Switch	1.0		
L_1 =		2.3 dB = 1.70		

$$\therefore \quad T_s = 110 + 290\,(\underbrace{1.70 \times 1.58 - 1}_{1.69}) = 110 + 488.9 = 598.9$$

$$(NF)_o = \frac{T_s}{T_o} = \frac{598.9}{290} = 2.07 = 3.15\text{ dB}$$

5. Determine System Losses —

L =	1. Atmospheric	2.8 dB	(Round Trip) (Ref. 3, Page 29)
	2. Beam Shape	3.2	(Ref. 2)
	3. Antenna — T_x	3.5	
	4. Antenna — R_x	2.8	(non-ohmic)
	5. Rotary Joint	0.1	
	6. Duplexer	0.5	
	7. Waveguide	0.5	
	8. Isolator	0.5	
	9. Harmonic Filter	0.1	
	10. Matched Filter	0.8	
L =		14.8 dB	

6. Establish interpulse period —

100 nm x 12.36 μs/nm	= 1236 μs	
retrace time	= 100 μs	(given)
t_o	= 1336 μs	

7. List remaining parameters for range equation —

$R^4 = (100\text{ nm})^4 = 10^8\text{ nm} = 80\text{ dBnm}$

$\sigma = 1.0\text{ m}^2 = 0\text{ dBm}^2$

$\nu = 0.28\text{ dB}$

8. Solve range equation for energy —

$$\mathcal{E} = \frac{S/N\ R^4\ L(NF)_o}{\nu\ G_T G_R\ \sigma\ \lambda^2}$$

S/N	= 23.40 dB	ν	= 0.28 dB
R^4	= 80.00	$G_T G_R$	= 93.00
L	= 14.80	σ	= 0
$(NF)_o$	= 3.15	λ	= 20.00
	121.35		113.28
	-113.28		

$\mathcal{E}$ = 8.07 dBj = 6.41 joules

9. Solve for peak power —

$$P_p = \frac{\mathcal{E}}{\tau} = \frac{6.41}{3.6 \times 10^{-5}} = \underline{178\ Kw \qquad answer}$$

10. Solve for average power —

$$P_A = \frac{\mathcal{E}}{t_o} = \frac{6.41}{1.336 \times 10^{-3}} = \underline{4.8\ Kw \qquad answer}$$

11. Solve for duty cycle —

$$\text{Duty cycle} = \frac{P_A}{P_p} = \frac{4.8}{178} = \underline{.027 \quad answer}$$

1.7 Summary — The radar range equation has been derived and reformatted for application to ESRS design. Single-pulse operation is stressed because it is characteristic of time-shared multifunction ESRS's. Other parameters emphasized in the use of the range equation for ESR application are:

- Antenna area gain
- Operating noise factor (with system noise temperature referred to antenna face)
- Ohmic antenna receive losses (included in the system noise temperature)
- Non-ohmic antenna receive losses (included in the system losses)
- Average power (how to determine for ESRS's employing variable energy, waveforms and interpulse periods)

References

[1] Kraus, J.D., *Antennas*, McGraw-Hill, 1950.

[2] Blake, L.V., "The Effective Number of Pulses per Beamwidth for a Scanning Radar," *IRE* pp. 770-774, June 1953.

[3] Blake, L.V., "Radar/Radio Tropospheric Absorption and Noise Temperature," *NRL Report # 7461*, October 30, 1972.

[4] DiFranco & Rubin, *Radar Detection*, Prentice Hall, 1968.

[5] Meyer & Mayer, *Radar Detection Theory*, Academic Press, 1973.

[6] Barton, D.K. & Ward, H.R., *Handbook of Radar Measurement*, Prentice Hall, 1969.

List of Symbols

A_o = antenna area — meters^2

A_R = physical area of receive aperture — meters^2

A_T = physical area of transmit aperture — meters^2

A_z = azimuth angular coordinate

B_N = noise bandwidth of the IF filter — hertz

D = aperture dimension — meters

D_A = azimuth dimension of antenna — meters

D_E = elevation dimension of antenna — meters

E = elevation angular coordinate

$\mathcal{E}$ = transmitter energy for single-pulse — joules

$\mathcal{E}_s$ = total energy in search mode — joules

F_1 = preamplifier noise figure

F_2 = mixer noise figure

G = antenna area gain

G_1 = preamplifier gain

G_2 = mixer conversion gain/loss

G_R = receive antenna area gain — $\dfrac{4\pi A_R}{\lambda^2}$

G_T = transmit antenna area gain — $\dfrac{4\pi A_T}{\lambda^2}$

K_R = fraction of the range sweep that the receiver is gated on ($K_R \approx 1$)

L = loss factor to account for system losses not included in system noise temperature losses

L_1 = ohmic receive antenna plus microwave receive losses

L_2 = losses between preamplifier & mixer

N = noise power — watts

n = number of beam positions required to cover search volume

$(NF)_o$ = operating noise factor

P_A = transmitter average power — watts

P_{A_s} = average power in search mode — watts

P_D = probability of detection

P_N = probability of false alarm

P_p = transmitter peak power — watts

R = range of target — meters

R_1, R_2 = two successive range measurements — meters

R_M = maximum detection range — meters

$\dot{R}_{12}$ = range rate from successive steps — meters/second

$\dot{R}_{12}, \dot{R}_{23}$ = successive range rates — meters/second

$\ddot{R}$ = acceleration from 3 successive steps — meters/second^2

S/N = r.f. signal-to-noise ratio of single pulse

T_o = reference temperature (usually 290° K)

T_A = antenna sky noise temperature — degrees Kelvin

T_s = system noise temperature referred to antenna face — degrees Kelvin

V = target radial velocity — meters/second

c = velocity of light = 3×10^8 — meters/second

f_D = doppler frequency — hertz

f_{D1}, f_{D2} = two successive doppler returns

$\dot{f}_D$ = acceleration from 2 successive doppler returns — hertz/second

f_r = pulse repetition frequency — hertz

k = Boltzman's constant = 1.38×10^{-23} watts/(hertz x degree Kelvin)

n = number of beam positions required to cover search volume

p(S,N)	=	probability density functions of signal-plus-noise and noise voltages
t	=	time taken by pulse to travel to target & return — seconds
t_o	=	interpulse period — seconds
t_{fa}	=	false alarm time — seconds
t_s	=	total frame time in search mode — seconds
θ_o	=	antenna 3 dB beamwidth
θ_A	=	azimuth 3 dB beamwidth
θ_E	=	elevation 3 dB beamwidth
λ	=	wavelength of transmission — meters or centimeters
ν	=	conversion factor — 1.06711 = .28 dB
ξ	=	scan angle from array normal
σ	=	radar target cross section — $meters^2$
τ	=	pulse width — seconds

Modern Radar Trends

Chapter Two

2.1 Historical Development — A brief summary of the major contributions to radar development is of historic interest and also indicates the trends of modern radar.

1886 - Hertz experimentally tested Maxwell's theories and demonstrated the similarity between electromagnetic (radar) and light waves. In particular, radio waves propagate with the velocity of light and are reflected by metallic and dielectric bodies.

1904 - Hülsmeyer, German engineer, obtained a patent for an obstacle detector and ship navigational aid.

1922 - Recommended use of short waves for radio detection.

1925 - Briet & Tuve, Carnegie Institute of Washington, used principle of pulse ranging which characterizes modern radar. It became the established method in all countries for measuring the height of the ionosphere.

1930s - Successful pulse radar systems were developed independently in the United States, England, France and Germany.

1940s - Modern radar began in the war years. Randell & Boot, England, developed pulsed magnetron thus raising radar frequencies from 300 to 3000 MHz region. High power was made possible with narrow beams and reasonable antenna sizes.

Stanford University developed the klystron which provides

higher power output and better frequency stability than the magnetron.

MTI, moving-target-indication, was introduced to obtain improved low altitude coverage. Since good MTI performance requires high frequency stability, the klystron replaced the magnetron in applications requiring good MTI performance.

Coherent radars, employing MOPA (master-oscillator-power-amplifier) chains were developed for improved clutter rejection and doppler frequency measurement (obtained by use of doppler filter banks or coherent memory filters).

1950s - More range capability and order of magnitude accuracy improvement were achieved to meet satellite, intercontinental ballistic missile and guided missile requirements.

Receiver sensitivity improved by an order of magnitude through the development of better crystal-mixers and the introduction of parametric amplifiers and solid state masers.

Very high power transmitters and large antennas were developed.

Closer integration of radar systems to weapons due to electronic computer techniques — aircraft fire control, SAGE (Semi Automatic Ground Environment), terminal defense — Nikes.

Frequency and frequency-phase scanning radars were developed mainly for naval applications.

1960s - Conventional radar improvements were mainly in the application of complex wide bandwidth waveforms and development of efficient signal processors. Analog and digital waveform generators, receivers/signal processors, and antennas were developed to operate with frequency coded and pulse burst waveforms of hundreds of megahertz bandwidth. Fast Fourier Transform (FFT), Time Compressed Spectrum Analyzer and Digital MTI processors were developed and successfully demonstrated in retrofitted operational radars.

This era experienced intensified activity in the ESR field.

High power, large aperture and high accuracy systems were developed.

ESRs became operational for satellite and ICBM detection.

ESRs were developed and undergoing evaluation in complex weapons systems for air and terminal defense.

Typical examples are AN/FPS-85 Space Track Radar, SAFE-GUARD sensors PAR (Perimeter Acquisition Radar) and MSR (Missile Site Radar), and SAM-D (Surface-to-Air Missile Development).

ESR studies and development programs were initiated for many other applications such as multifunction airborne radars, mortar and weapon locators, and air traffic control. Solid State ESR designs were studied and developed in the latter part of the decade. In particular: transmit/receive modules, phase shifters, and mixers.

1970s - Effort on the major ESR programs has continued into the early 1970s. Many of the ESR studies of the 1960s have become development programs in the 1970s. In addition, more sophisticated wide signal bandwidth long-range ESR systems are being built for operational evaluation.

2.2 **Modern Radar Requirements** — In the last decade the growing need for radars to provide increased range coverage, target identification, trajectory determination, faster data rates (short reaction time), and accommodate increased target densities; to cope with high speed jet propelled aircraft, guided missiles, ballistic missiles, and satellites has imposed new technological and system design requirements.

Increased Range Coverage —

Rewriting the energy equation (1.27) with

$$G_T\ G_R = G^2 = \left[\frac{4\pi A}{\lambda^2}\right]^2 \qquad (1.10)$$

obtains

$$R^4 = \frac{K'\ E\ \sigma\ A^2}{S/N\ L\ NF\ \lambda^2} \qquad (2.1)$$

that clearly indicates the controllable design parameters affecting range performance.

Design for:

1. *Decreased noise power density*

Use low noise preamplifiers such as:
transistors, TDAs (tunnel diode amplifiers), TWTs (travel-

ing wave tubes), uncooled or cooled parametric amplifiers and masers.

The preamplifier selected in a particular radar application is dependent on performance specifications, "balanced design" based on system trade-off studies, and cost effectiveness.

2. *High transmission energy*

Obtained with high peak power and/or long time duration coded waveforms.

The maximum peak power obtainable with a single tube is approaching a practical design barrier of approximately 50 megawatts using klystrons.

The maximum pulse time duration is restrained by transmitter limitations, minimum range performance specifications, and compatibility of time-bandwidth factors with range resolution (signal bandwidth) requirements.

3. *Large antenna aperture/high antenna gain*

Practical construction considerations and manufacturing tolerances limit attainable beamwidths to approximately 0.1°.

Corresponding antenna gain is approximately 58 dB depending on the illumination taper. At an operating frequency of 1.0 GHz the aperture is in the order of 600 feet diameter.

4. *Minimized system losses*

Locate transmitter and receiver preamplifier as close as feasible to antenna feed to reduce waveguide and cable losses.

Obtain maximum signal processing efficiency by using matched filter waveform processing for: single tone pulses, frequency coded pulses, pulse bursts, MTI and pulse doppler waveforms. Employ correlator devices like FFT - Fast Fourier Transform or Time Compressed Spectrum Analyzer for good signal processing efficiency.

Employ pulse-to-pulse frequency diversity to minimize post detection integration losses in applications requiring high probabilities of detection on fluctuating targets.

Pulse-to-pulse frequency diversity also reduces power-aperture requirements in cases requiring high cumulative probabilities of detection with only 3 or 4 looks on fluctuating targets.

Increased Resolution — High range and angular resolution provide improved accuracy since accuracy is inversely proportional to signal bandwidth and directly proportional to beamwidth. Higher accuracy leads directly to improved trajectory determination for locating launch and impact points of missile targets and satellite orbit determination. High resolution enables target identification by providing length measurement and identification of the major scatterers of various body shapes. Another desirable feature of high resolution is the improved capability of detecting targets in clutter. Less clutter is intercepted by the radar since smaller space cell increments are interrogated on each look. If the resolution is equivalent to signal bandwidths of several hundred megahertz, the radar can "see" in-between the clutter patches and detect targets without MTI or pulse doppler processing.

Design for:

1. *Wide instantaneous bandwidth systems in the order of several hundreds megahertz*

   All components, antenna, transmitter, receiver, signal processor, display, data processor, and even interconnecting cables and connectors must be capable of wide bandwidth operation. Extreme care is required to avoid or eliminate multiple reflections, differential delays, and transients.

2. *Narrow pencil beams*

   Narrow antenna beams require low antenna sidelobes for proper radar operation.

   Near-in targets and/or targets with large radar cross sections (50 to 60 dBm^2) located at sidelobe angular positions will provide echo signals that exceed the detection threshold and be reported erroneously to be in the angular position corresponding to the main beam direction unless the one-way sidelobes are significantly lower than 30 dBs.

   To obtain very low sidelobes in the order of 40 dB requires precise antenna construction to tight tolerances using a high illumination taper.

Shorter Reaction Time — Very short reaction time or almost instantaneous radar reporting of target presence is required in many modern radar applications. Typical examples are: to allow time to catalog satellites while they are rapidly passing through the radar coverage volume; to discriminate ICBM clouds and take appropriate counteraction against the reentry vehicles prior to impact; and to identify a large number of fast moving targets and to assign and guide missiles to destroy them in complex weapon systems.

Design for:

1. *High data rate*

   The radar system must be capable of both gathering and processing large amounts of data very rapidly.

2. *Multifunction*

   The radar system must be integrated to perform all the required functions of search, multitarget acquisition, multitarget track, multimissile track, multimissile guidance, and kill evaluation simultaneously or on a time shared basis.

3. *Automatic Operation*

   Automatic computer programming of multimode radar operation and processing of data is a necessity in a high data rate system providing very short reaction time.

The logical goal in radar development is to seek a system design satisfying all the radar requirements in one integrated system.

ESRs have the inherent capability to meet this goal.

ESRS Character-istics

Chapter Three

3.1 Definition — Electronic scanning is defined as a method of positioning an electromagnetic beam in space by electronic means with the antenna aperture remaining fixed and no mechanical mechanism involved in the scanning process.

For comparison, the definition of mechanical scanning is a method of positioning an electromagnetic beam in space by a mechanical rotation or angular positioning of the radiating aperture of the antenna system.

There are also compromise methods. The most common employs one dimension electronic scanning in elevation and mechanical rotation in azimuth.

3.2 Electronic Scanning Techniques — In the basic configuration of the radar (Fig. 3.1) the exciter/transmitter generates the signal waveform to be transmitted. The phasing system subdivides the transmitter power into a number, N, of equal (or weighted) packets, and appropriately phase codes each packet to form a wavefront (beam) in the designated direction. The N signals are amplified in the transmit modules (omitted in some designs) and radiated by the antenna elements. The N echo return signals from the target are preamplified in the receive modules (omitted in some designs) and coherently added together in the phasing system to obtain a composite target echo signal that is processed by a conventional receiving system before being displayed. The computer programs the operation of the radar and processes the data.

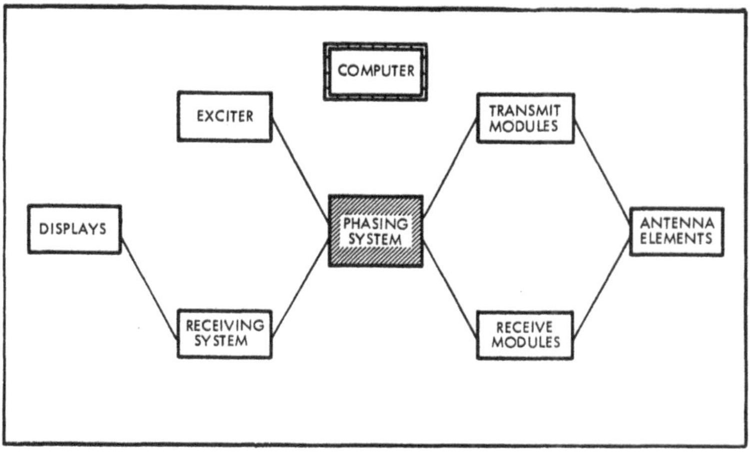

Figure 3-1 Basic Elements of an Electronic Scan Radar. ESR Radar Systems Differ Primarily in the Phasing Systems Used to Steer the Antenna Beam. The Computer Connects to all Other Parts of the System.

The essential differences among the various methods of implementing an electronic scanning radar (ESR) occur in the phasing system.

The selection of a phasing system, or electronic scanning technique, to meet a potential user's performance requirements is one of the major design decisions to be made by the system architect.

The basic electronic scanning techniques are:

Phase,
Real-Time,
Frequency, and
Electronic Feed Switching.

Phase — In a phase scanning system, electronic beam position scanning is accomplished by varying the phase of the separate elements of the array in a predetermined manner. This operation is illustrated in Fig. 3.2 for a linear array. A number of radiating elements are equally spaced along a straight line. Feeding these elements with in-phase currents of equal or tapered amplitude results in the propagation of a wave front parallel to the array, i.e., the formation of a broadside beam. As indicated in Fig. 3.2, the direction in which the beam is formed is an arc sine function of the interelement differential phase shift. A unique beam position is obtained for each progressive phase shift setting in the network feeding the array.

Continuous phase scanning is accomplished by exciting all radiating elements coherently with a small frequency increment between elements as indicated in Fig. 3.3. The beam is continuously scanned at a rate equal to the frequency differential f.

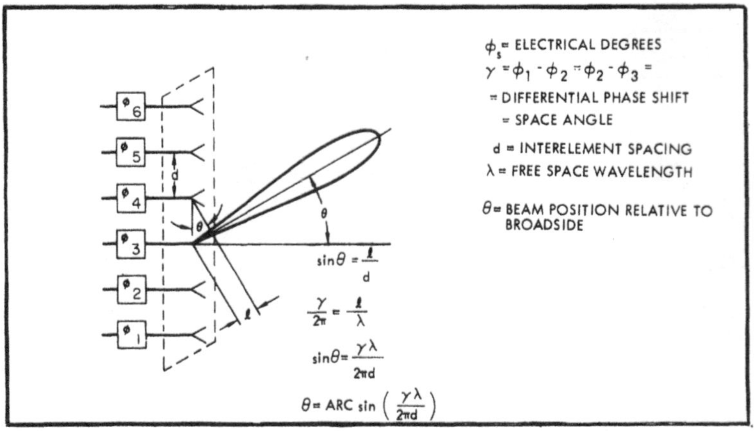

Figure 3-2 Basic Operation of a Phase Scanning System. Note that Beam Direction Depends on Phase Shift Between Elements.

Phase shifters employed for phase scanning can be reciprocal and nonreciprocal, analog and digital, ferrite and diode. Reciprocal phase shifters introduce the same phase regardless of which way the power passes through them. Thus the same setting can be used for receiving as for transmitting. This has far reaching implications. Many modern radars use pulse-burst transmission rather than a single pulse per repetition interval. Typically, a burst contains about 100 subpulses spread over a millisecond. Because of the burst length, it may be necessary to receive between subpulses. With reciprocal phase shifters this can be done at the same setting, since it is necessary to reset only to steer to a new beam position. Since nonreciprocal phase shifters must be reset twice for every subpulse, in this typical application the nonreciprocal device must be adjusted about 200 times more often than the reciprocal.

Phase shifters are also classified as analog and digital. The analog device produces a phase shift proportional to an applied voltage or current. The digital phase shifter might better have been called a binary phase shifter. It produces phase shifts in an incremental binary fashion (180°, 90°, 45°, 22.5°, etc.) upon command. It is well suited to working directly with digital computers — therefore, the name digital phase shifter.

Two widely used materials for phase shifters are semiconductor diodes (usually the PIN type) and ferrites (usually yttrium-iron-garnet). Diode phase shifters to date have all been of the reciprocal digital type. A three-bit digital phase shifter is shown in Fig. 3.4. Using ferrites, both reciprocal and nonreciprocal phase shifters in either analog or digital versions are possible (Fig. 3.5). Diode devices of today are favored over ferrites at lower frequencies because of lower r-f losses. Ferrite phase shifters are favored at high frequencies because the magnetic losses are less. The losses are comparable at approximately 5 GHz.

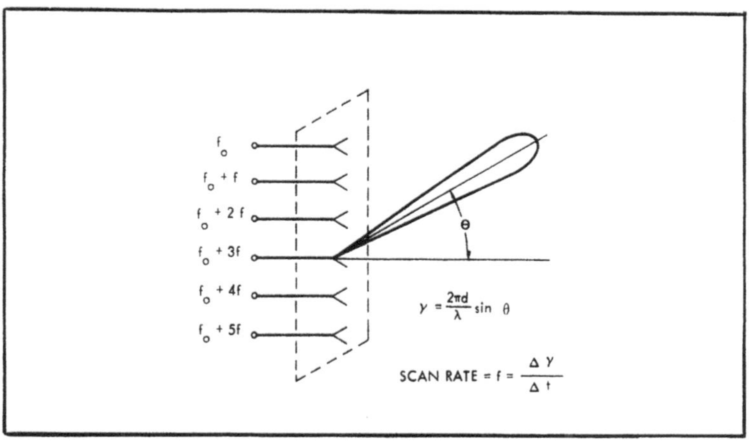

Figure 3-3 Continuous Phase Scanning

Real Time — Real time scanning is a special case of phase scanning. Instead of employing phase delay to form a wavefront (beam) in a designated direction, time delay is used. The difference is demonstrated in Fig. 3.6. Time delay is required in ESR systems having wide (instantaneous) signal bandwidths. The relationship among signal bandwidth, beamwidth, transmission frequency and maximum scan angle is derived in Appendix A and given in Eq. (3.1):

$$B \leqslant \frac{\theta_o f_o}{128 \sin \theta_m} \text{ (deg)} \tag{3.1}$$

Phase scanning may be used if Eq. (3.1) is satisfied. For example if $\theta_o = 1^\circ$, $f_o = 3$ GHz and $\theta_m = 60^\circ$, the maximum signal bandwidth that can be used with phase scanning is approximately 27 MHz.

In some applications a compromise design is feasible employing a combination of phase and time delay devices.

Time delay devices consist of different lengths of cable being switched in-and-out in binary increments in a similar manner to the operation of diode phase shifters.

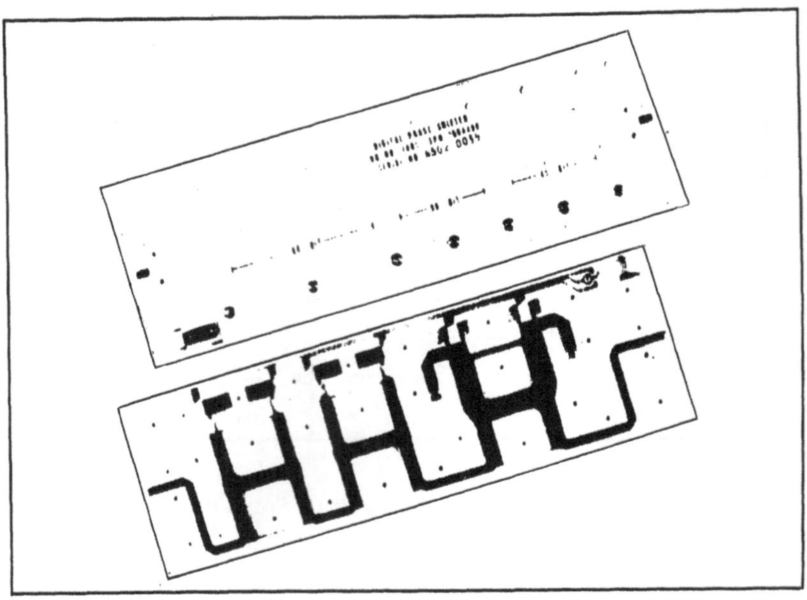

Figure 3-4 Three-Bit Diode Phase Shifter

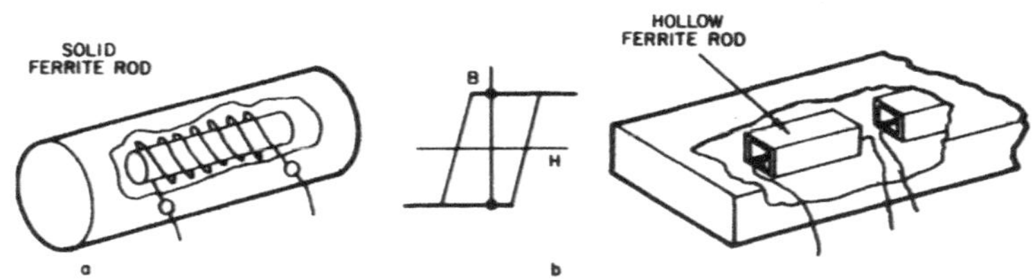

A reciprocal analog ferrite phase shifter (a) and a nonreciprocal digital ferrite phase shifter (b) are illustrated. In (a), the solenoid provides longitudinal magnetization of the ferrite rod. This device theoretically can give large phase shift per unit length of ferrite thus minimizing r-f losses in the ferrite. Temperature sensitivity is severe, 1 deg phase shift per 1 deg C change, thus requiring a close control of the array temperature.

It suffers from hysteresis effects and needs a fairly large continuous magnetizing current. The digital phaser (b) uses ferrite having a square hysteresis loop, as shown in the B-H curve. Here a single turn provides the magnetization for each of the hollow rods or bits. The ferrite is driven into saturation, and the current is then removed. (Since only two operating states are used, the device is sometimes called a latching phase shifter.) Temperature sensitivity is somewhat better, about 0.1 deg phase shift per 1 deg C temperature change. The hysteresis problem is vastly simplified, and continuous current is not required.

Typical switching speeds are in the 1 to 10 μsec region. Research on reciprocal types of ferrite phase shifters is being pursued actively.

Figure 3-5 Ferrite Phase Shifter

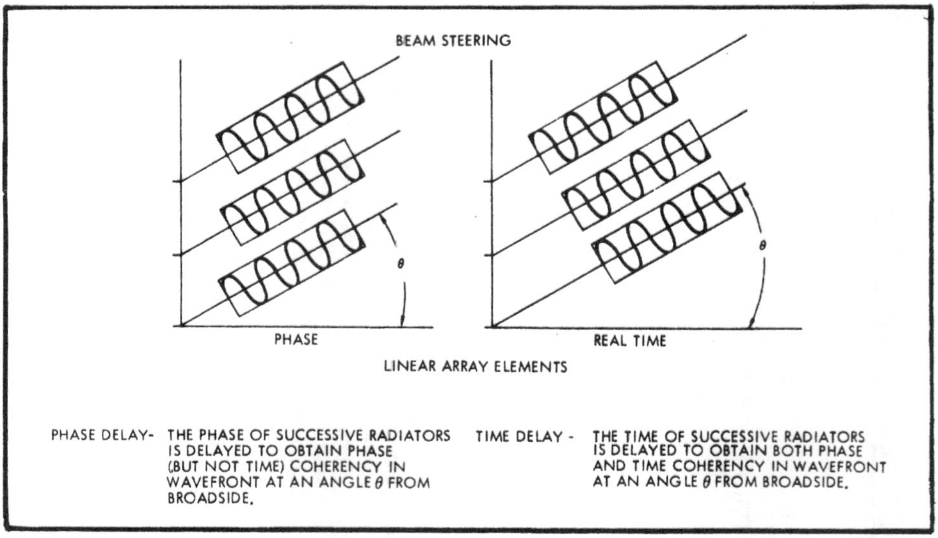

Figure 3-6 Comparison of Phase and Real Time Scanning

Frequency — In a frequency scanning system, frequency is employed to control the interelement differential phase shift so that each frequency corresponds to a unique beam position. This method is completely passive and reciprocal and is illustrated in Fig. 3.7.

Multiple beams can be formed simultaneously by transmitting a sequency of adjacent signal waveforms, each at a different center frequency; where each center frequency corresponds to a designated beam position. Multiple beam reception is obtained by connecting multiple receivers to the receiver output of the duplexer; each receiver tuned to receive one of the transmitted frequencies.

Practical designs employ approximately a 6% bandwidth to scan 90 degrees. Either helical (shown in Fig. 3.7) or serpentine feeds are employed in frequency scanning radars (FSR's). The helical feed can operate over a wider frequency range than the serpentine feed. The relationship among signal bandwidth, beamwidth, frequency scanning bandwidth, and maximum scan angle is derived in Appendix B and given in Eq. (3.2):

$$B \leqslant \frac{\theta_o \Delta f}{256 \sin \theta_m} \quad \text{(deg)} \tag{3.2}$$

Frequency scanning may be used if Eq. (3.2) is satisfied. For example, for the same values used in the phase scanning example, i.e., $\theta_o = 1^\circ$, $f_o = 3$ GHz, $\therefore \Delta f = 180$ MHz (.06 x 3 GHz) and $\theta_m = 60^\circ$, the maximum signal bandwidth that can be used with frequency scanning is approximately 0.8 MHz (compared to 27 MHz for phase scanning).

Electronic Feed Switching — Two simple examples of electronic feed switching are illustrated in Figs. 3.8 and 3.9. In both techniques multiple beams can be formed simultaneously. The hybrid/Butler matrix shown in Fig. 3.8 is a special case of phase scanning whereas the Luneburg lens shown in Fig. 3.9 is a real time (wide signal bandwidth) scanning technique.

The basic concept of the hybrid phasing matrix can be explained by referring to the four-element array shown in Fig. 3.8. The system is completely passive and reciprocal, and consists of an interconnection of hybrids, and either fixed phasing elements or differing lengths of strip transmission line.

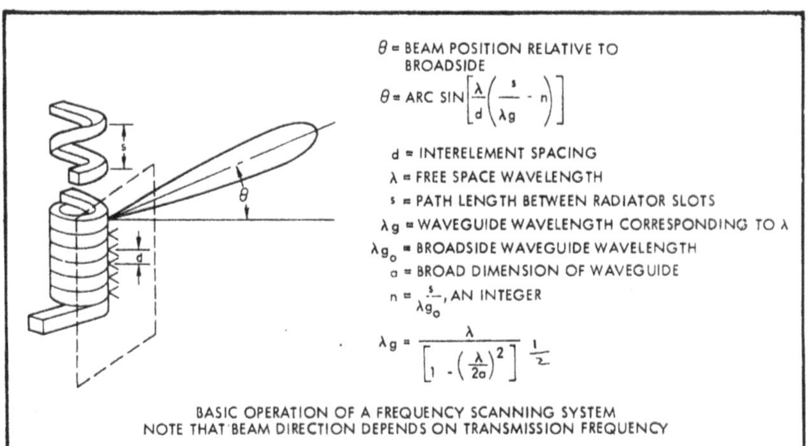

Figure 3-7 Basic Operation of a Frequency Scanning System. Note That Beam Direction Depends on Transmission Frequency.

The beams are formed in accordance with the interferometer equation for a linear array:

$$\Delta\phi = \frac{2\pi d}{\lambda} \sin\theta \tag{3.3}$$

where

$\Delta\phi$ = interelement phase shift
d = interelement spacing
λ = free-space wavelength
θ = beam position relative to broadside

An input connection, transmit or receive, to any of the four input ports results in the formation of the corresponding antenna beam. Formation of Beam 2 (in Fig. 3.8) requires a differential phase shift of 45 degrees between elements. It corresponds to excitation of the antenna elements in the sequence zero degrees, 45 degrees, 90 degrees, and 135 degrees. The intrinsic phase difference of 90 degrees between the outputs of a 3-dB directional coupler (hybrid) provides the proper phase differential for alternate antenna elements. The fixed phase shifter converts the 90 degree difference between the outputs of the input hybrids to a 45 degree differential in order to excite the two groups of alternate antenna elements properly.

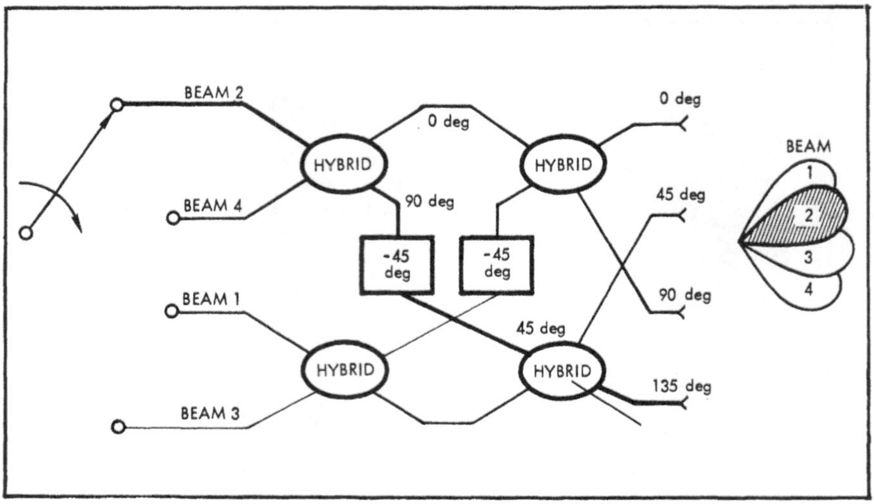

Figure 3-8 Hybrid/Butler Matrix Basic Operation of an Electronic Feed Switching System

The hybrid phasing matrix for a more complex antenna also consists of a structure of hybrids and either fixed phase shifters or differing lengths of strip transmission line. The matrix for an antenna with N elements has N input ports to establish N beams in space; in all instances, it is a repetitive structure of hybrids and fixed phasing elements.

An orthogonal set of (sinx) /x beams is formed crossing at the 4.0 dB point with 13.5 dB maximum sidelobes. Cosine taper with 23 dB maximum sidelobes can be obtained by coherently combining adjacent input ports. Coherent combination with proper weighting of three input ports to the hybrid phasing matrix results in a cosine-squared taper with 32 dB maximum sidelobes. In theory, the coverage sector should approach 180 degrees as the number of radiating elements increases. In practice, the coverage sector is limited by the beamwidth of the individual radiating elements. In most applications, strip transmission line construction can be used.

A geodesic lens may also be used for electronic feed switching. One example is the Luneburg lens shown in Fig. 3.9. A geodesic line is the shortest superficial line between two points on any surface. According to Fermat's principle of least time, this is the path along which energy will propagate between these two points. A geodesic lens is a surface that satisfies these conditions; such a surface is designed by methods of differential geometry. A geodesic lens is inherently a wide bandwidth device.

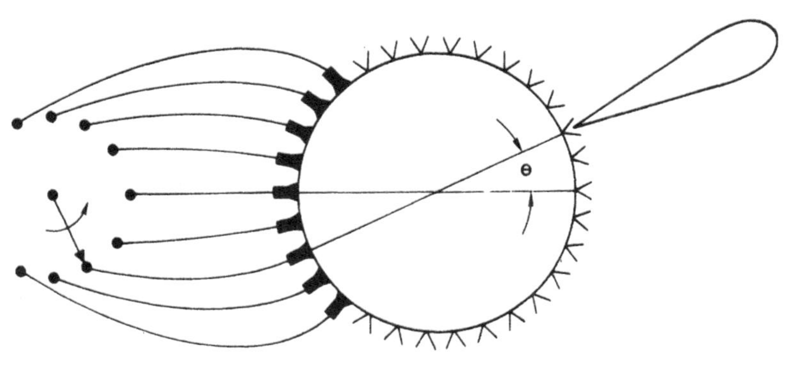

Luneburg Lens

Figure 3-9 Basic Operation of an Alternative Electronic Feed Switching System

In both types of electronic feed switching, individual transmitters can be added at each aperture in order to provide power amplification and permit low-power electrical switching. In Fig. 3.8 the switch determines which part of the phasing matrix receives or transmits the energy. This selection, in turn, determines the direction of the beam formation. Similarly, the Luneburg lens shown in Fig. 3.9 is employed to obtain the correct phasing to propagate a wave front perpendicular to the axis of the feed being activated.

3.3 **Implementation** — How the basic four techniques are employed, individually or in combination, to implement ESR system designs varies widely. Typical one and two dimensional configurations are described in this section employing line, cylindrical, planar and spherical arrays to form electronically scanned beams. Conformal arrays (where the surface of the array conforms to the shape of, say an aircraft or a missile) are special cases of the above.

One Dimension — A simplified one-dimensional phase scanning radar system is indicated in Fig. 3.10. A linear array is employed to scan the beam electronically in elevation. Three methods are shown to focus the beam in the unscanned azimuth dimension. The parabolic reflector is the first method used. This was replaced by the more efficient planar or billboard array which consists of horizontal rows of elements fed in-phase by strip line corporate feeds. The differential phase shift between elevation elements is obtained in this example by the use of 3-bit (180°, 90°, 45°) diode phase shifters. The phase setting determining the elevation beam direction is controlled by the scan programmer. Azimuth coverage is obtained by mechanical rotation of the antenna. The programer settings can be made during the retrace time between transmitted pulses so that no additional time is required to slew the antenna beam to any desired position within the elevation scan sector. The programer is either a digital-analog or completely digital device. Conventional transmitters and receivers are used in this system.

A basic one-dimensional frequency scanning system is shown in Fig. 3.11. Frequency scanning was actually the first technique to be fully implemented in an operational radar. Two methods of forming the pencil beam to obtain one-dimensional frequency scanning are shown in Fig. 3.11. As the transmitted frequency is varied in accordance with the scan program, the interelement phasing will vary and radar beam steering will result. The radiating elements may be either primary slots or dipole radiators. Aperture amplitude taper (by control of each radiating element) may be employed to reduce sidelobe levels. Azimuth coverage is obtained by mechanical rotation of the antenna.

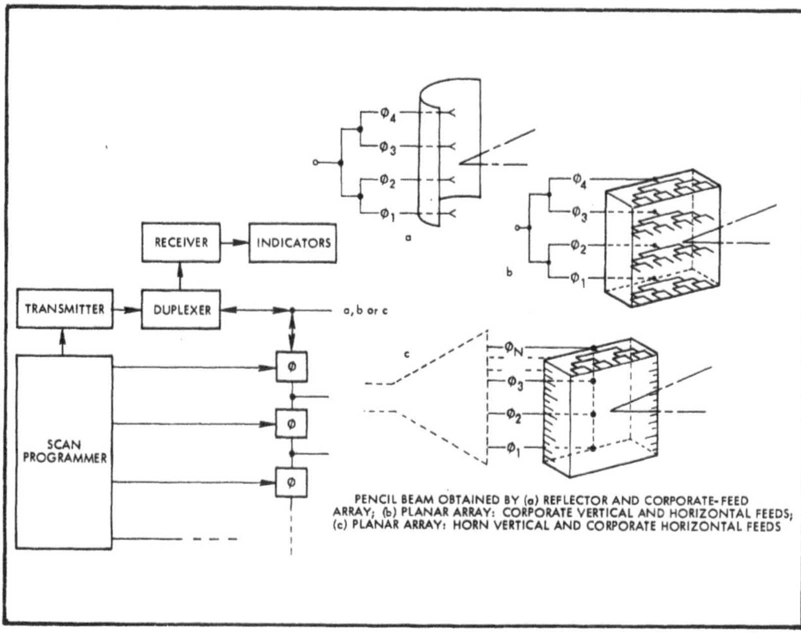

Figure 3-10 One-Dimensional Phase Scanning Radar System

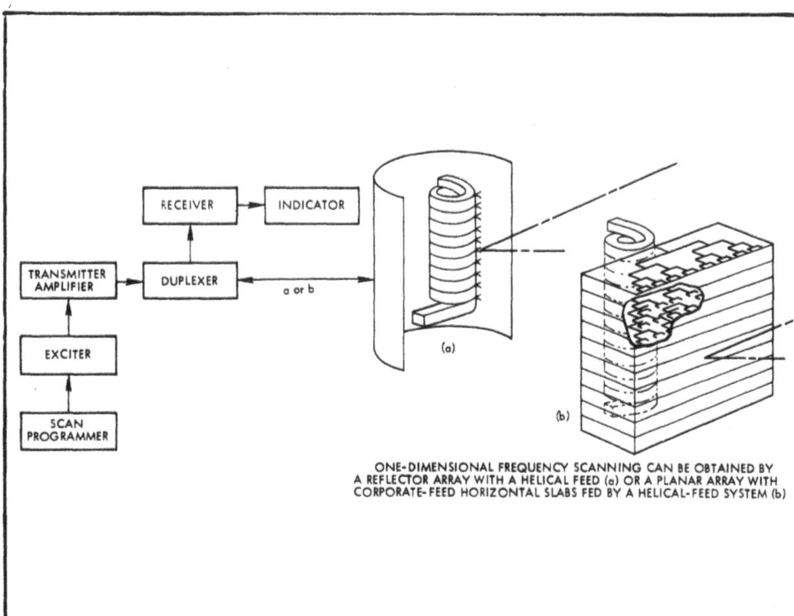

Figure 3-11 One-Dimensional Frequency Scanning Radar System

The receiver system is quite conventional. The exciter-transmitter is usually of a MOPA (Master Oscillator Power Amplifier) design using crystal oscillators to obtain the required frequency stability. The frequency is changed in discrete steps to obtain the various beam positions by appropriate gating, mixing, filtering and frequency multiplication of the energy obtained from a bank of crystal oscillators. The scan programer provides the required gates to accomplish the beam slewing during the retrace time.

A one dimensional electronic feed switching scanning system can be designed in the form of a circular cylinder consisting of a stack of Luneburg disks. Vertical corporate feeds supply in-phase energy to the disks. Azimuth electronic scanning is obtained by electronically switching vertical feeds that are sequentially transmitting and receiving the energy. Multiple beams in azimuth can be obtained simultaneously by using multiple transmitters and receivers feeding multiple vertical feeds.

Another form of a one-dimensional electronic feed switching scanning system can be designed using a hybrid phasing matrix in the vertical plane in place of the vertical corporate feed and phase shifters shown in Fig. 3.10a, b. In this design elevation electronic scanning is obtainable by electronically switching input terminals. As in the previous case, multiple inputs can be fed simultaneously to obtain a stacked beam radar.

Two Dimensions — Various combinations and modifications of the one-dimensional electronic scanning techniques can be made to obtain two-dimensional scanning. In applications where wide bandwidth waveforms are required, real-time scanning techniques are employed such as various types of geodesic lens or replacement of phase delays by time delays. Frequency scanning is not applicable to wide bandwidth applications.

One means of obtaining two-dimensional electronic scanning is illustrated in the functional diagram of Fig. 3.12 where frequency scanning is employed in both dimensions. The elevation frequency determines the elevation beam position indirectly by supplying the differential phase shift, ϕ_E, by means of a delay network. The elevation scanning frequency is eliminated by the mixing process so that only the phase shift appears at the antenna, superimposed on the azimuth scanning frequency, f_A. The antenna consists of a stack of helical or folded(serpentine) waveguides in which the azimuth frequency determines the beam position in the azimuth plane and the differential phase shift between the folded waveguides determines the beam position in the elevation plane. The azimuth and elevation frequencies

—This two-dimensional system uses frequency scanning in azimuth; the elevation frequency contributes a phase component which is used for positioning elevation scanning beam.

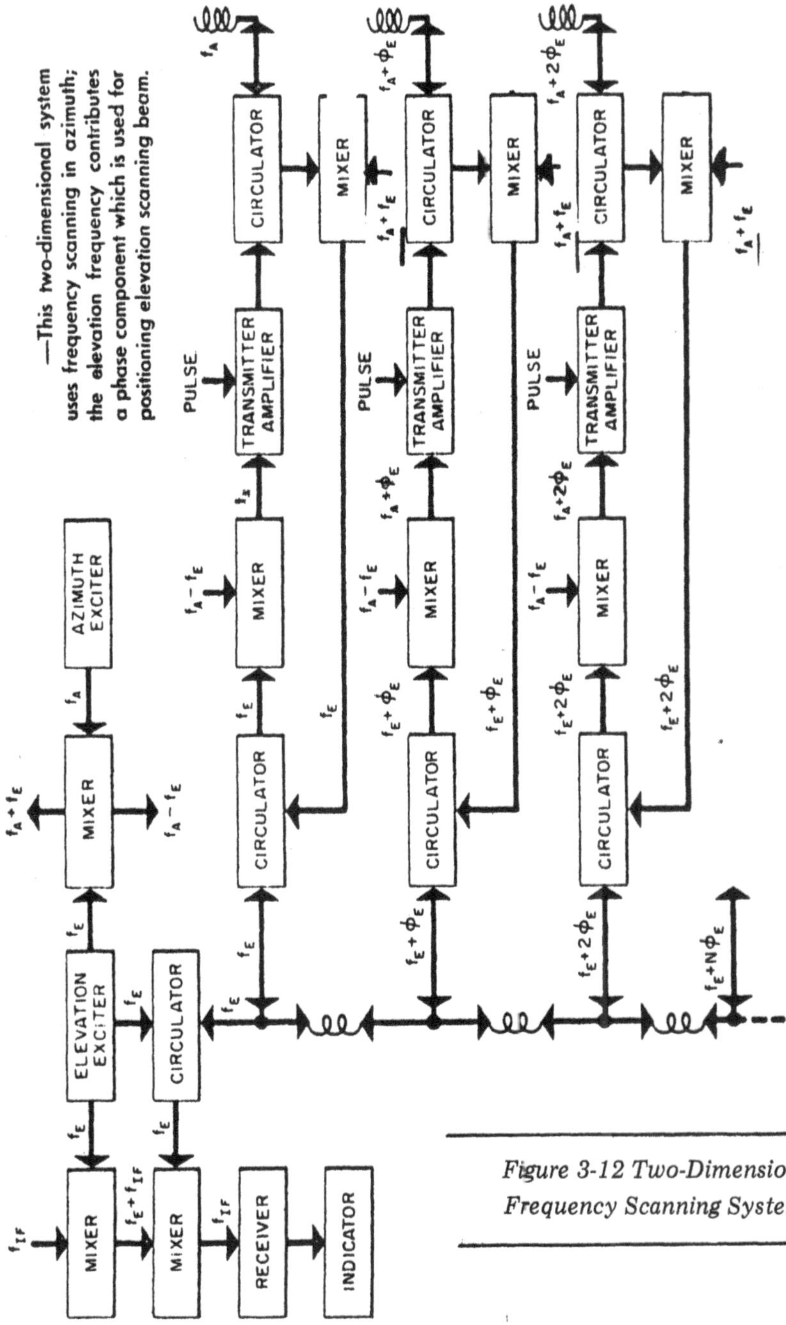

Figure 3-12 Two-Dimensional Frequency Scanning System

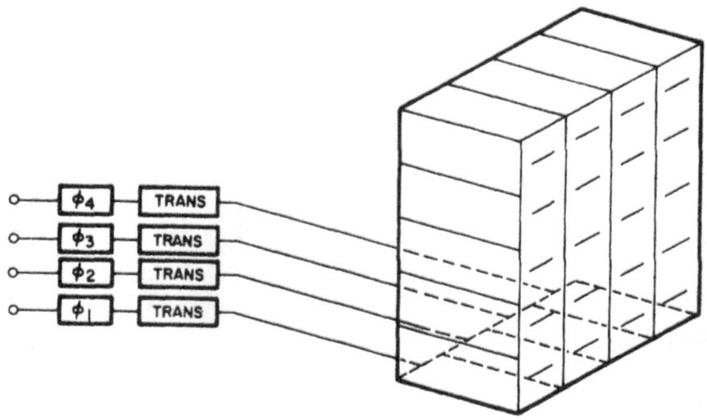

the antenna is a planar array having columns of serpentine waveguides, each one having an individual power amplifier. The frequency is changed in increments to obtain step scanning in elevation. The azimuth scan is obtained by using a phase shifter at the input of each transmitter to control the differential phase shift between the serpentine columns.

Figure 3-13 Two-Dimensional Frequency/Phase Scanning System

can be programmed independently to a first order approximation to obtain beam position slewing in either or both directions. The exterior of the antenna has the appearance of a flat billboard. This type of system can scan 90 degrees in elevation and 90 degrees in azimuth without any mechanical motion of the antenna. Four of these systems will provide complete hemispherical coverage with an electronically scanned beam.

Another simple variation employs frequency scanning in elevation and phase scanning in azimuth (Fig. 3.13). The antenna is again a planar array consisting of columns of folded waveguides, each one having an individual power amplifier. The frequency is changed in increments to obtain step scanning in elevation. The azimuth scan is obtained by using a phase shifter at the input of each transmitter to control the differential phase shift between the serpentine columns.

The next logical step is to employ a lens having a cellular structure with a ferrite or diode digital r-f phase shifter in each cell (Fig. 3.14). By combining the phase increment associated with the vertical and horizontal beam positioning in each module, two-dimensional scanning is obtained. This is known as an optical feed or space feed array.

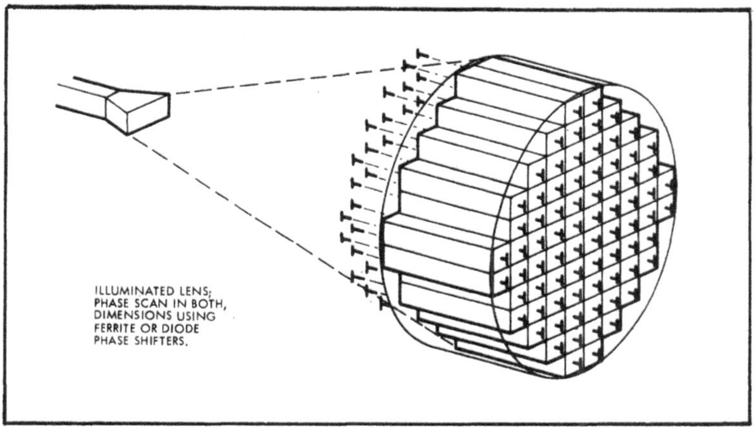

Figure 3-14 Two-Dimensional Space Feed Array

Two-dimensional phase scanning is also obtained by employing a two-dimensional corporate feed input to a cellular structure with a ferrite or diode r-f phase shifter in each cell as shown in Fig. 3.15. This corporate or confined feed array can be transformed to a wide signal bandwidth design by supplementing the phase shifters with real time delay devices.

Two-dimensional electronic feed switching is obtained by cross stacking one-dimensional strip-line hybrid phasing matrices. In Fig. 3.16, a 16 element array is illustrated which employs eight strip-line hybrid phasing matrices. Sixteen orthogonal beams are formed that may be scanned sequentially or interrogated simultaneously. Individual transmitters can be added at each element for power amplification and to permit switching at low power.

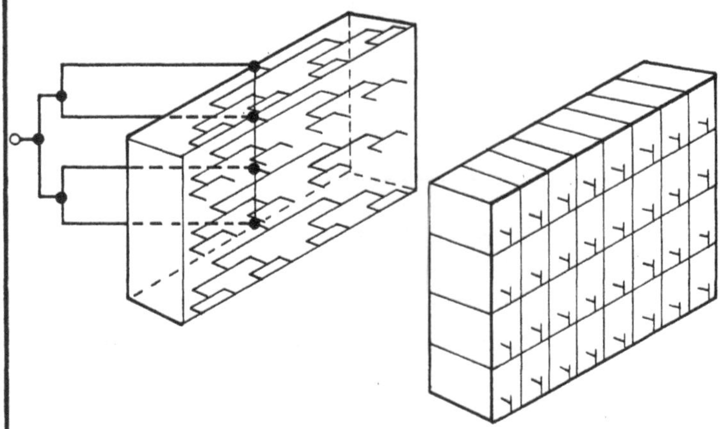

Figure 3-15 Two-Dimensional Confined Feed Array

An alternate two-dimensional electronic feed switching system is obtained by employing a spherical Luneburg lens as a phasing system to switch-scan electronically over a complete hemisphere (Fig. 3.17). It is inherently a wide signal bandwidth device. This device can also form orthogonal beams that may be scanned sequentially or interrogated simultaneously. Individual transmitters can be added at each element for power amplification and to permit switching at low power.

The most popular array for ESR is the planar configuration. In its construction and in ease of electrical design, it has distinct advantages over curved shapes. Lower sidelobes can be achieved because of its superior aperture illumination. Maintenance of a flat structure is simpler than a curved one. Its chief restriction is a limit of 60-75 degrees scan angle off broadside so that 3 or 4 faces are required for full hemispherical coverage. On the other hand, curved arrays, such as a sphere, can use only a portion of the radiating elements for any one beam direction because of the inability of elements on the back side to radiate in a forward direction. In addition, planar arrays make more efficient use of the full aperture.

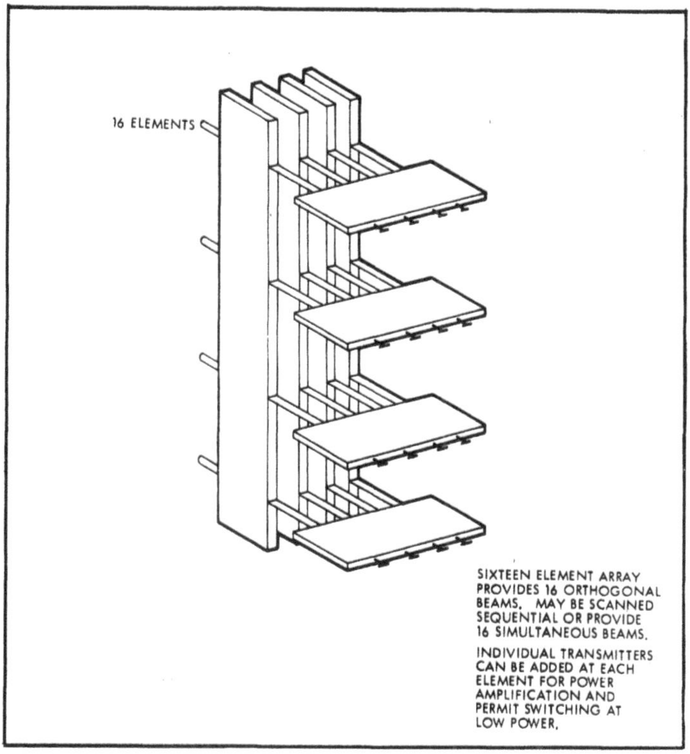

Figure 3-16 Two-Dimensional Hybrid Phasing Matrix Array

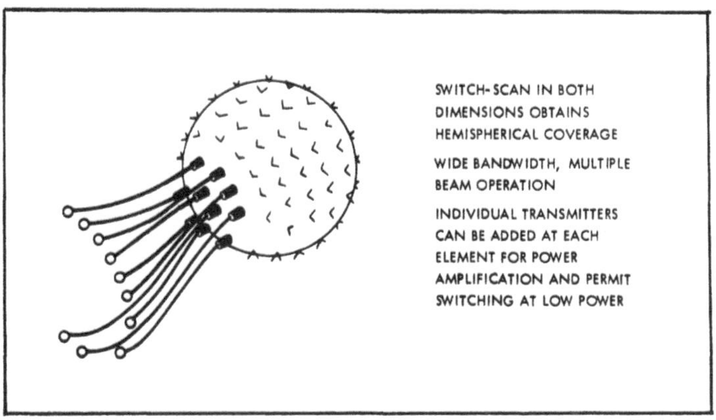

Figure 3-17 Two-Dimensional Spherical Luneburg Lens Array

Beam Splitting Techniques — In a multipurpose radar, consideration must be given to the tracking technique and the effects of multitarget tracking on tracking performance. ESRS's (in particular, single beam designs) are characterized by relatively infrequent sampling of individual targets compared to conventional tracking radars in order to be able to track say, up to 100 targets on a time-shared basis. Sampling rates are in the order of 0.5 to 5.0 looks per second on individual targets compared to 100 to several thousand looks per second of conventional tracking radars. Consequently, beam splitting techniques are favored that do not require excessive sampling rates on individual targets.

The four most common beam splitting techniques are:

Amplitude Weighting,
Sequential Lobing,
Amplitude Monopulse (Simultaneous Lobing),
Phase Monopulse.

Amplitude weighting is the simplest and least accurate beam splitting technique. The intensities of the target echoes in a beam are monitored and the angular position is determined by some form of greatest of selection process. This beam splitting technique is not recommended for most ESRS applications because of the relatively poor accuracy and the relatively large number of looks (5 to 10) required per beam position on individual targets.

Figure 3-18 One-Dimensional, Confined Feed, Phase Monopulse Phase Scanning System

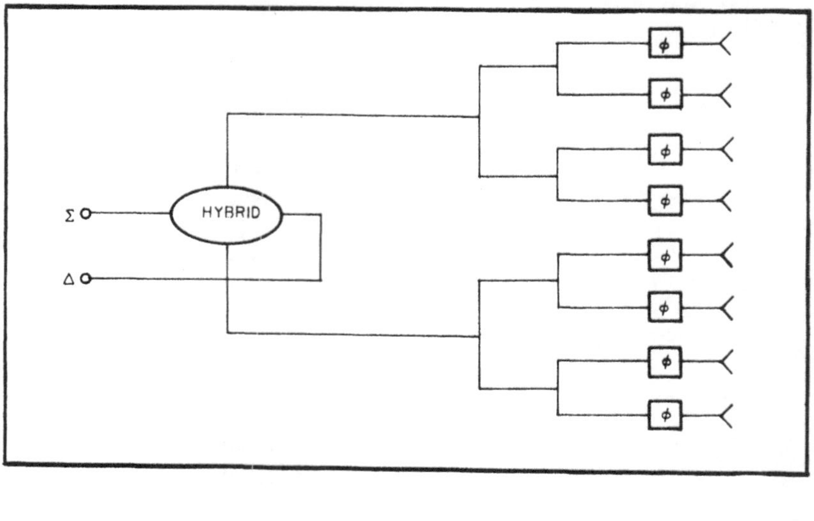

Sequential lobing is easy to mechanize in all types of ESRS's. It is adequate for some applications but has the disadvantages of introducing target amplitude scintillation as an additional source of tracking error (worse in a frequency scanning system) and also reduces the subclutter visibility in a MTI (Moving-Target-Indicator) or a pulse doppler system.

Amplitude monopulse is the most accurate beam splitting technique with minimum ambiguities and sidelobe levels. A phase monopulse beam splitting technique is more sensitive to random phasing errors than amplitude monopulse because it employs four identical phasing systems for two-dimensional beam splitting with one-fourth of the radiating elements involved in minimizing the rms phasing error compared to an amplitude monopulse system. This discrepancy is less significant in arrays consisting of a large number (4000 to 40,000) of active radiating elements.

Since monopulse beam splitting provides angular position data on a single hit basis, it is very suitable for ESRS applications. The degree of antenna complexity associated with implementing monopulse beam splitting is dependent on the ESR design as will be indicated by describing several types.

1. Confined Feed, Phase Monopulse

Phase monopulse beam splitting in a confined feed ESR is achieved by conventional interferometer techniques. This is illustrated in Fig. 3.18 for a one-dimensional phase (or real time) scanning system. The two halves of the linear array are summed and differenced in a microwave hybrid or rat race. The same principle is extended to a two-dimensional application as shown in Fig. 3.19. In essence, there are four separate corporate feed structures feeding a cellular planar array containing a phase shifter for each active radiating element.

2. Confined Feed, Amplitude Monopulse

Amplitude monopulse beam splitting in a confined feed ESR is relatively complicated to implement. A microwave hybrid is required between adjacent radiating elements as shown in Fig. 3.20 for a one-dimensional phase (or real time) scanning system. Another implementation employs a hybrid between elements half an aperture length apart. Many variations of this design are possible. A two-dimensional amplitude monopulse beam splitting system (Fig. 3.21) is obtained by duplicating the one-dimension configuration for every horizontal row of active elements and adding three vertical corporate feeds with an additional hybrid for each pair of adjacent horizontal

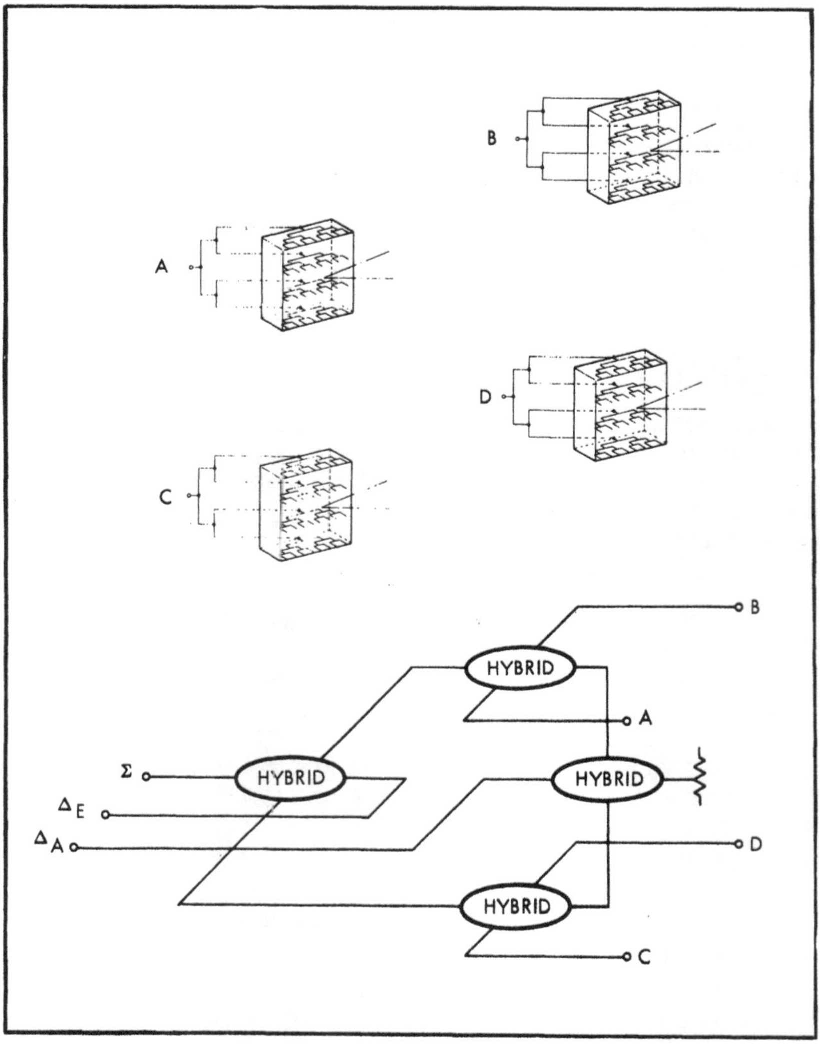

Figure 3-19 Two-Dimensional, Confined Feed, Phase Monopulse Phase Scanning System

rows to obtain the elevation difference channel. Therefore, the equivalent of three corporate feed networks are required to obtain amplitude monopulse beam splitting. The confined feed has the advantage of providing equal power distribution to each active element in design applications where a transmitter module is used at each radiating element, with the flexibility of having intermediate power amplifiers interlaced within the feed structure to avoid use of a single high-power exciter source. In addition, it is relatively easy to design the feed for multiple squinted beams if they are required.

In some applications the two methods are combined to provide phase monopulse in one angular dimension and amplitude monopulse in the other.

3. Space Feed, Amplitude Monopulse

The space fed phased array (Fig. 3.14) has the advantage of providing amplitude monopulse beam splitting by the use of a conventional monopulse horn. Either a 5 horn or 4 horn multimode configuration can be used to obtain amplitude monopulse beam splitting in much the same manner as a conventional tracking radar.

4. Hybrid Phasing Matrix, Amplitude Monopulse

Continuous beam steering for amplitude monopulse beam splitting is achievable by proportioning power between adjacent beams of the hybrid phasing matrix. A method for obtaining analog motion of a beam is illustrated in Fig. 3.22. Power is supplied to two adjacent ports of the hybrid phasing matrix simultaneously. If all the power is supplied to Port 1, Beam 1 is formed; if all the power is supplied to Port 2, Beam 2 is formed; if equal power is supplied to both ports, a beam is formed half way between Beams 1 and 2. As the power ratio to the two ports is varied, the resulting antenna beam moves smoothly between the positions of the two beams. The power divider, Fig. 3.22, consists of two hybrids and a variable microwave phase shifter. Varying the phase from zero to 180 degrees moves the resulting antenna beam smoothly from Position 1 to Position 2.

Amplitude monopulse beam splitting is achieved by duplicating the circuitry shown in Fig. 3.22, and connecting it between Ports 2 and 3 of the hybrid phasing matrix. Both variable phase shifters are synchronously controlled as shown in Fig. 3.23. This provides smooth motion of the null plane of a pair of beams for amplitude monopulse beam splitting.

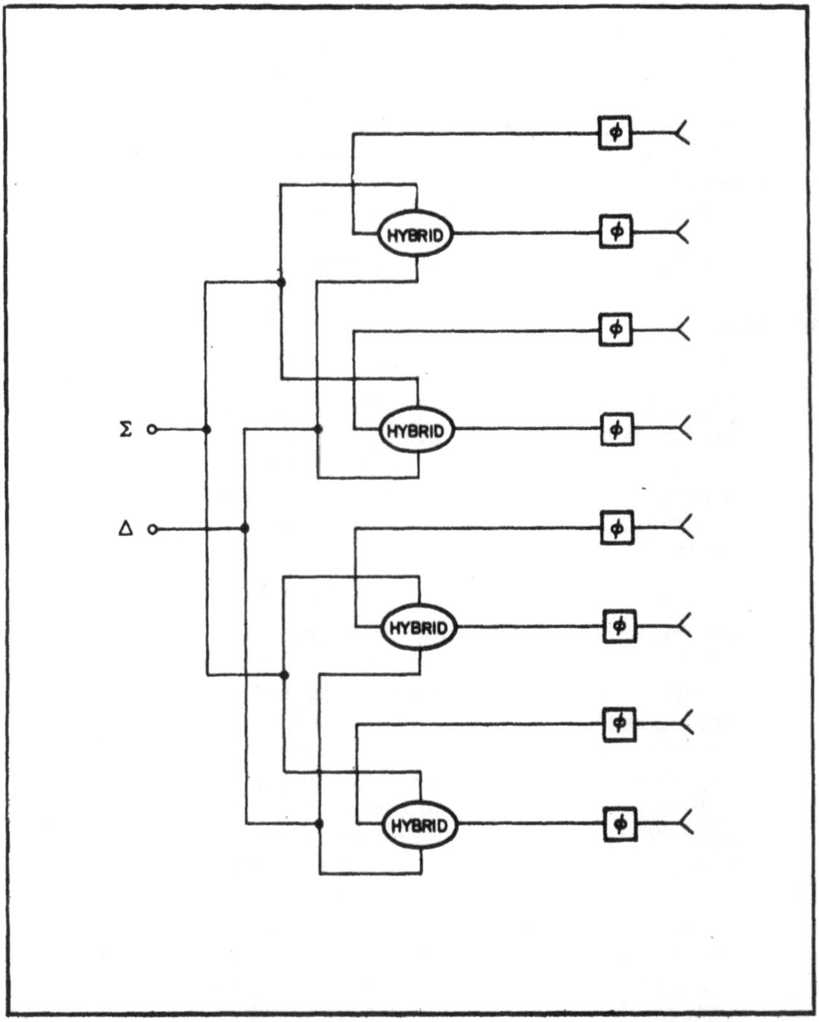

Figure 3-20 One-Dimensional, Confined Feed, Amplitude Monopulse Phase Scanning System

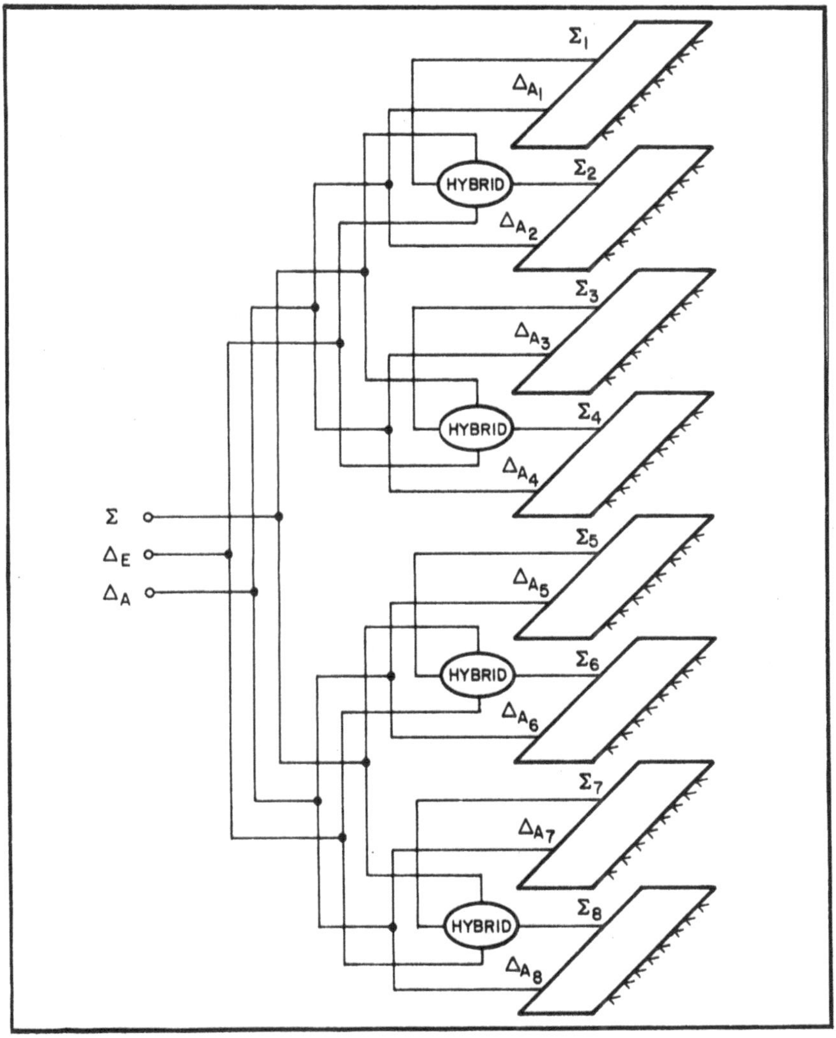

Figure 3-21 Two-Dimensional, Confined Feed, Amplitude Monopulse Phase Scanning System

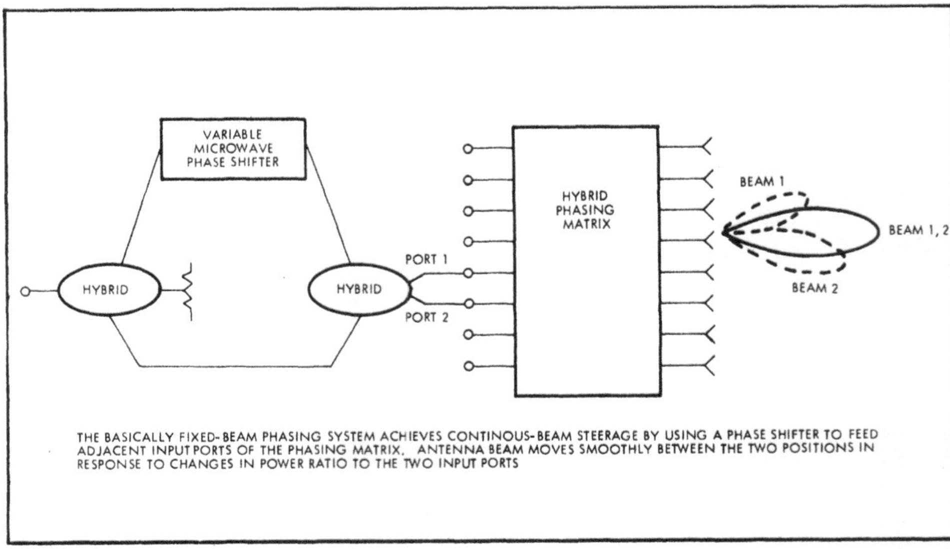

Figure 3-22 Hybrid Phasing Matrix — Continuous Beam Steering

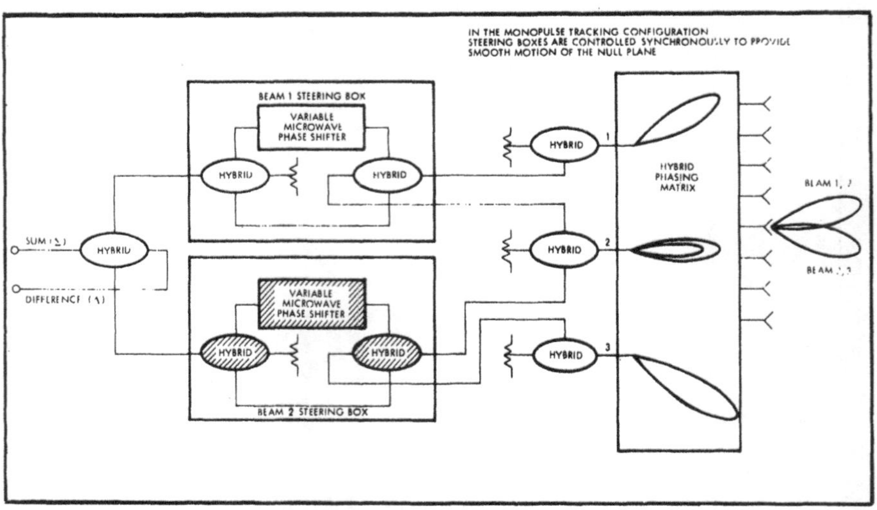

Figure 3-23 Hybrid Phasing Matrix — Amplitude Monopulse Beam Splitting

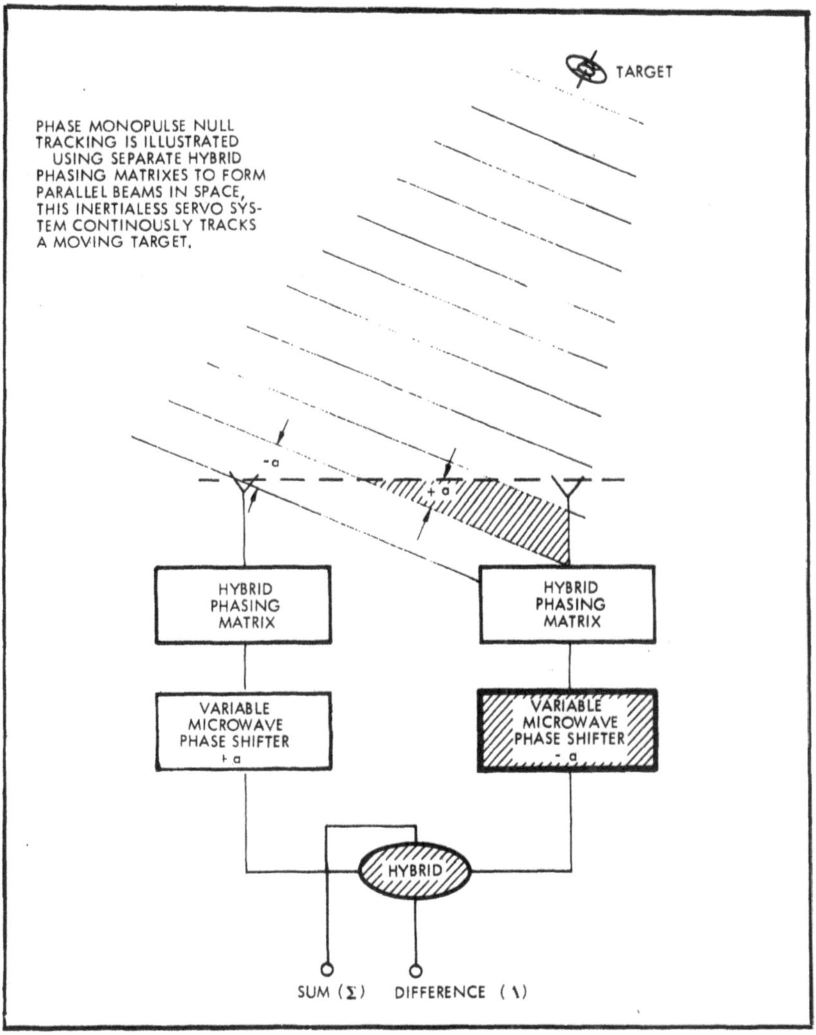

Figure 3-24 Hybrid Phasing Matrix — Phase Monopulse Beam Splitting

5. Hybrid Phasing Matrix, Phase Monopulse

The variable microwave phase shifter can be used to obtain phase monopulse beam splitting by conventional interferometer techniques as shown in Fig. 3.24. Separate hybrid phasing matrices form parallel beams in space directed at the target. The angular target location is determined by the phase differential between the target return signals. The difference is caused by the lengths of the paths from the target to the antenna. The error signal obtained from this phase differential is amplified and used to control the variable microwave phase shifters. Thus an inertialess servo system is formed that can continuously track a moving target.

The amplitude/phase monopulse steering box (Fig. 3.25, p. 67) permits a combination of amplitude and phase monopulse null tracking. The upper hybrid phasing matrix is connected to the upper half of a planar array and the lower hybrid phasing matrix is connected to the lower half of the array. The steering box has six output terminals. The upper three are attached to adjacent horizontal input ports of the upper matrix to obtain amplitude monopulse null tracking. The lower three are attached to the corresponding ports of the lower matrix to perform the same function. Amplitude monopulse sum-and-difference azimuth signals are obtained from the upper and lower comparators. These signals are processed through the four variable phase shifters and the remaining two comparators to obtain the total coherent sum signal, elevation phase monopulse difference signal, and azimuth amplitude monopulse difference signal. The two difference signals are amplified and used to control the corresponding variable microwave shifters. This results in a two-dimensional inertialess servo system that continuously tracks a moving target.

Monopulse implementation of geodesic lens electronic feed switching systems can be achieved in a similar manner to hybrid phasing matrices.

Illumination Tapering — Phased arrays have made it possible to obtain better tapering of the illumination pattern to minimize sidelobes. In the past designers had only the primary horn pattern to work with and usually settled for a cosine-on-a-pedestal taper providing 23 dB maximum sidelobe levels. Other tapers such as the Taylor with designs providing 40 dB maximum sidelobe levels are applicable to arrays. By individual control of a large number of radiating elements on rigid stationary array structures it is possible to design for and achieve antenna patterns with 40 dB maximum sidelobe levels.

Two basic methods are amplitude taper and space taper. In the amplitude taper, the energy fed to each element is diminished toward the edges of the antenna to approximate the desired pattern. The objection to this method is that more than half the elements are working at low power. Thus cost and weight penalties are incurred for equipment not fully used. Phased arrays, uniquely within the class of radar antennas, permit space taper. In the space tapering method, every active element carries full power, but instead of active elements being located at every radiating point, some are passive. Thus the active elements are less dense toward the outer edge of the antenna. The net effect on the antenna pattern is the same, but for identically sized arrays the space-tapered array has fewer active elements than the amplitude-tapered array. Space tapering is also called density tapering. The basic technique is known as thinning. Approximately 50 percent thinning is employed in thinned array designs where half the antenna elements are active and half are passive. The gain for a thinned array is proportionately less than for a "fully-filled" array. Fifty percent thinning reduces the aperture gain by 3 dB.

In some applications it is advantageous to employ a combination of amplitude and space taper. A space feed array is especially suitable for this combination since the feed horn can be amplitude-tapered and the array can be space-tapered.

Independent taper design is sometimes employed on transmit patterns and receive patterns to obtain maximum energy on target on transmission with relatively poor sidelobe levels and minimum sidelobe levels on reception. It is especially desirable to have independent transmit and receive illumination tapers in systems employing active transmit modules on each radiating element because the transmitters are operated at saturation levels and consequently even illumination or space tapering is employed for transmission. The receive taper is usually very high to obtain very low sidelobe levels to compensate for the poor transmit sidelobe levels. It is preferable to use amplitude taper on reception. It is accomplished with attenuators without loss of sensitivity provided there is adequate preamplification .

In monopulse operation, it is additionally desirable to taper independently the sum and difference illuminations in order to minimize the sidelobe level response on both. An intolerable sidelobe problem exists in phase monopulse operation if the sum and difference illuminations are not independently tapered.

3.4 **Advantages of ESRS** — In the previous chapter it was indicated that modern radar requirements are:

Increased range coverage,

Increased resolution,

Shorter reaction time,

and the logical goal in radar development is to seek a system design to satisfy all the radar requirements in one integrated system. Among the advantages of ESRS is the inherent capability to meet this goal and satisfy the above requirements, through the following features:

1. *Simultaneous Multimode Operation* —

   Search
   Multitarget Acquisition
   Multitarget Track
   Multimissile Guidance
   Automatic Reacquisition of Lost Targets
   Kill Evaluation
   Passive Detection

2. *Large Power-Aperture Products* —

   Very high power transmission capability (one transmitter per radiating element).

   Large fixed antenna arrays - high angular resolution (large structures can be rigidly constructed since no mechanical motion is required).

3. *Fast Reaction Time* —

   High data rates are obtainable because of multiple simultaneous beam capability and inertialess slewing of antenna beam position.

4. *Automatic Operation* —

   ESRS s are particularly suitable for computer control and data processing, thus providing rapid, efficient, and accurate radar programming and data processing. (Also contributes to fast reaction time).

5. *Electronic Stabilization* —

   Space coordinates are obtained referenced to a stable element thus eliminating the requirement for a heavy stable pedestal for moving platform application.

6. *Reliability* —

   Array component redundancy (active elements, phase shifters, transmit/receive modules) minimizes probability of catastrophic failure and provides graceful degradation.

7. Maintainability —

Computer controlled self-test diagnostic routines pinpoint trouble areas rapidly and minimize down-time for repairs.

Array component redundancy reduces logistics demands by requiring fewer different types of spare components to be delivered or stored.

8. Growth Potential —

In ESRS designs using multiple simultaneous beams, transmitters and receivers can be added at a later date to increase data rate and obtain faster reaction time. Larger computers or additions to computers (Central Processing Units, CPUs, and/or memories) can be added to handle the growth system.

In ESRS designs, larger computers or additions to computers can be added to increase multifunction and/or multitarget track capabilities.

In ESRS thinned array designs, passive elements can be replaced with active elements (and transmit/receive modules).

3.5 Disadvantages of ESRS —

1. Complexity —

The increased capabilities and flexibility of the ESRS also leads to more complex designs.

2. Cost —

More complex designs are more expensive.

3. Development of New Components —

Arrays	
Transmitters	
Special Purpose Computers	— beam steering units
Phasing Systems	— frequency scan feeds, phase shifters, real time delay devices, geodesic lens.
Monopulse Receivers	— with single hit normalization or computer normalization.
Consoles and Displays	— different and additional controls and displays.

The decision of whether to use conventional radars or ESRS s is different in each specific application and is dependent on a cost-effectiveness analysis and risk analysis. If both approaches achieve the same performance, the one which is least expensive and more likely to be delivered on schedule with minimum risk is the system to be selected. In some applications weight and size are also important considerations.

The more that is required of a radar system (multifunction, high data rate, handle large number of targets) the more likely it is that ESRS designs will be recommended over conventional radars.

3.6 **Examples of ESRSs** — Figs. 3.26 - 3.44 are examples of the various types of ESRSs that have been developed.

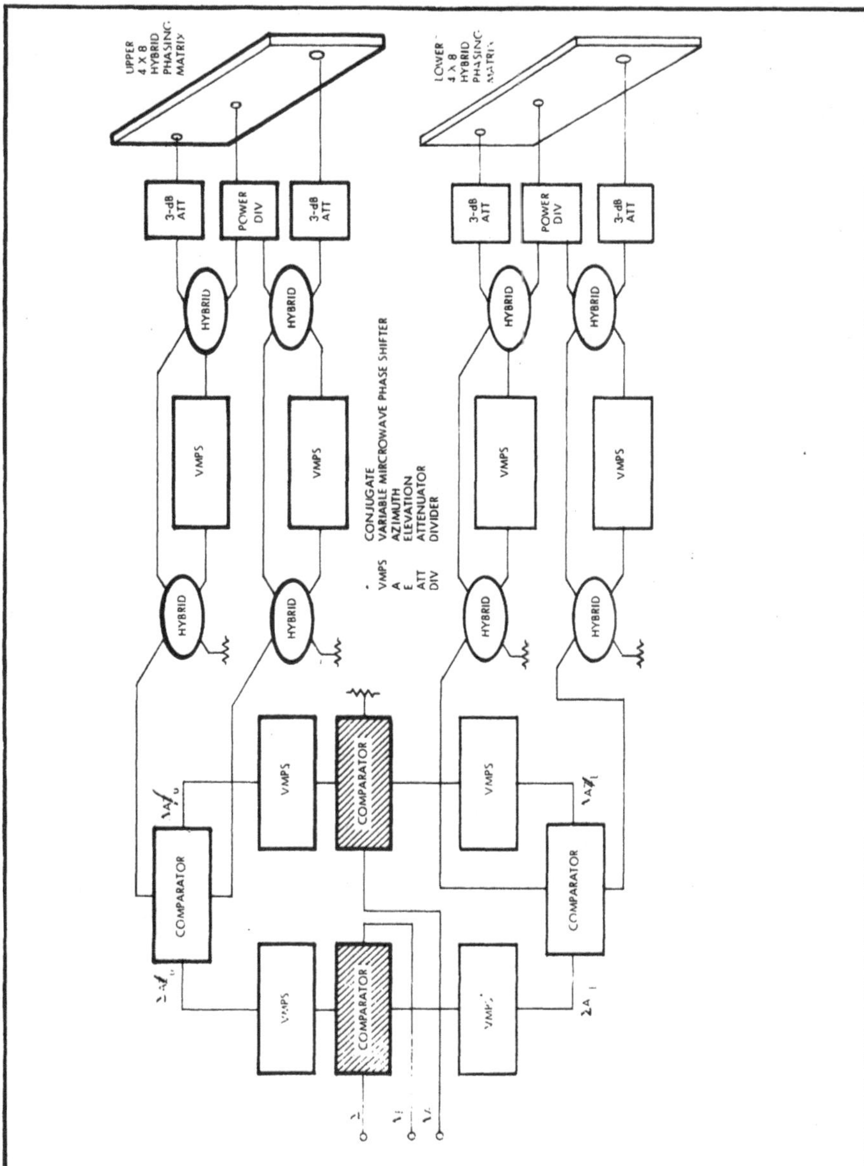

Figure 3-25 Hybrid Phasing Matrix — Amplitude/Phase Monopulse Beam Splitting

Fig 3-26 Safeguard Missile Site Radar — MSR
Meck Island Test Range — Kwajalein
Optical Feed — Fully filled array
Courtesy of Raytheon Company

Fig 3-28 Surface-to-Air Missile Development — SAM-D Minimum Fire Unit
Optical Feed — Fully filled array
Multifunction Radar — Search/Track/Guidance
Courtesy of Raytheon Company

Fig 3-27 Safeguard Missile Radar — MSR cutaway
Courtesy of Raytheon Company

Fig 3-29 AN/TPN-19 Landing System
Optical Feed — Reflect array
Courtesy of Raytheon Company

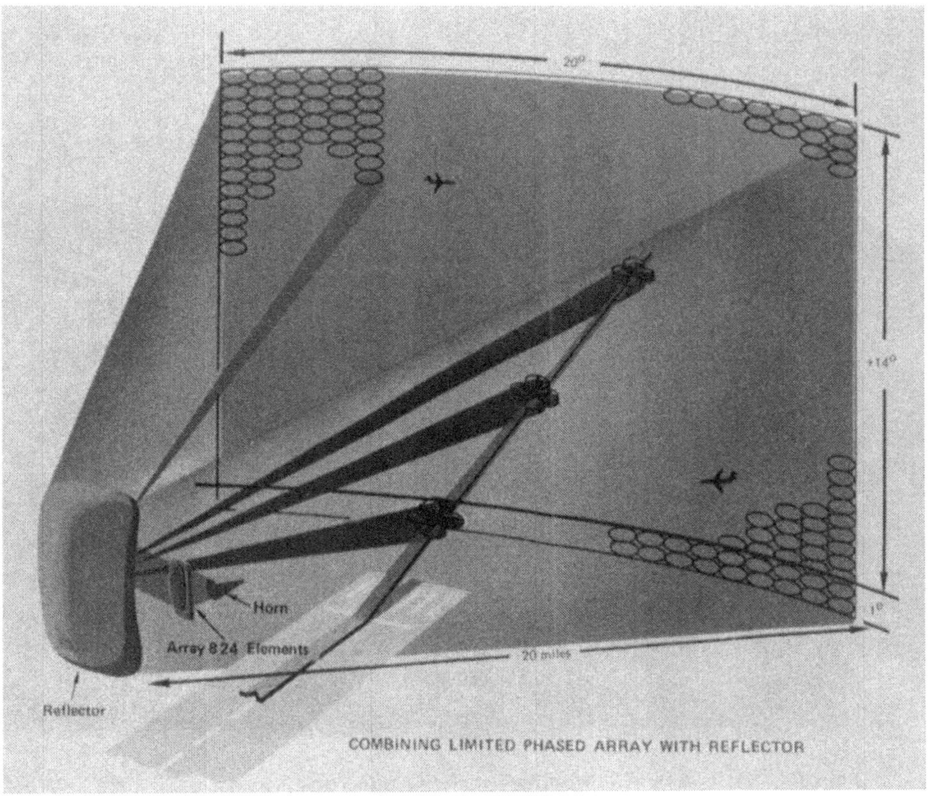

Fig 3-30 AN/TPN-19 Array with Reflector
Limited scan — 1/10 number of active elements
Courtesy of Raytheon Company

Fig 3-31 COBRA DANE — Intelligence Data Collector
Shemya AFB, Alaska
Confined feed — thinned array
Courtesy of Raytheon Company

Fig 3-33 Safeguard Perimeter Acquisition Radar — PAR cutaway
Courtesy of General Electric Company

Fig 3-32 Safeguard Perimeter Acquisition Radar — PAR
Confined feed array
Courtesy of General Electric Company

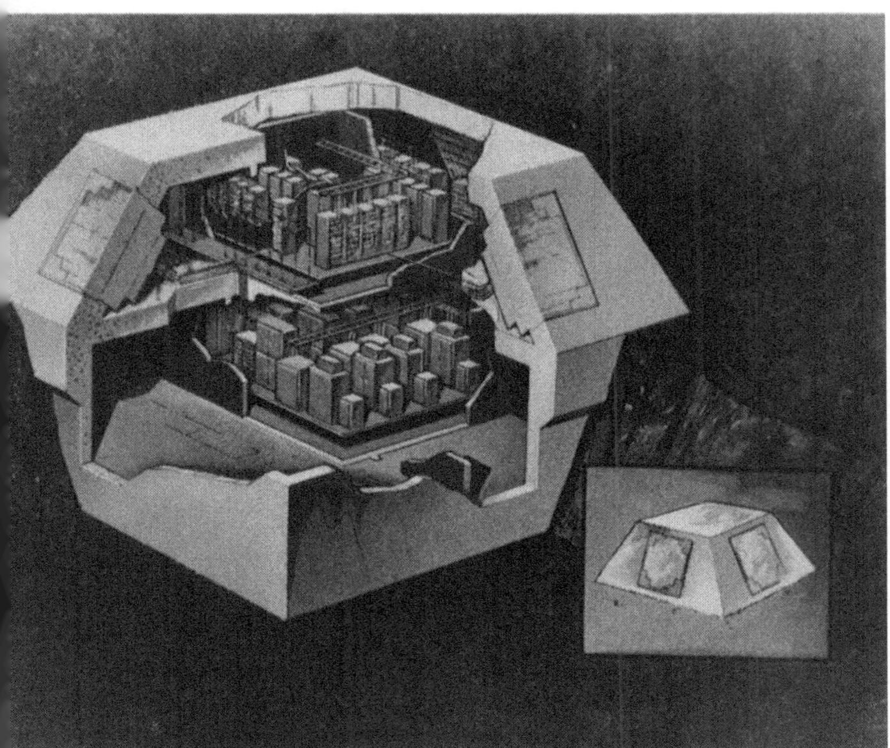

Fig 3-34 AN/TPS-59 3-D Phased Array Tactical Radar
Confined feed — solid state transmitter
Azimuth-Mechanical Rotation, Elevation — Phase Scan
Courtesy of General Electric Company

Fig 3-35 Electronically Steered Array Radar —ESAR
Demonstration Model
Courtesy of Bendix Corporation

Fig 3-36 AN/FPS-85 Space Detection & Track — SPADAT
Separate transmit & receive arrays
Courtesy of Bendix Corporation

Fig 3-38 AN/SPS-52 d-D Frequency Scanar Radar
Azimuth-Mechanical rotation, elevation — frequency
Courtesy of Hughes Aircraft Company

Fig 3-37 AN/SPS-32, 33 Search & Track Radars
USS Enterprise
AN/SPS-32 — Frequency scan
AN/SPS-33 — Frequency/phase scan
Courtesy of Hughes Aircraft Company

Fig 3-39 Hardpoint Demonstration Array Radar — HAPDAR Optical Feed — Thinned Aperture Computed Lens — TACOL Courtesy of Sperry-Rand Corporation

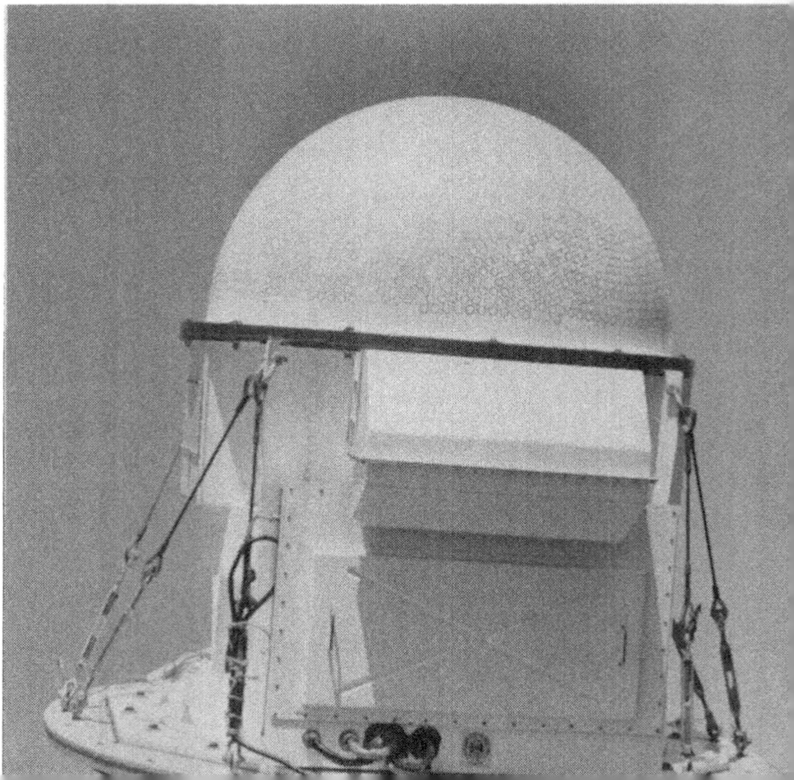

Fig 3-41 AN/SPY-1 Phased Array Antenna
Part of Advanced Electronic Guided Intercept System — AEGIS
Courtesy of Radio Corporation of America

Fig 3-40 Wide Angle Scanning Array Lens — WASCAL
Hemispherical coverage with single dome antenna
Courtesy of Sperry-Rand Corporation

Fig 3-43 Multi-function Array Radar — MAR-I
Demonstration model — White Sands Missile Range
Courtesy of GTE Sylvania

Fig 3-42 AN/SPS-48 Air Surveillance Radar
Azimuth — mechanical rotation
Elevation — Multi-beam frequency scan
Courtesy of ITT Gilfillan

Fig 3-44 Typhon — Multibeam Luneburg Lens.
Demonstration model
Courtesy of John Hopkins University, Applied Physics Laboratory and U.S. Navy

Appendix A

Build-Up Time in a Phase Scan Array

D = length of array aperture

θ_m = maximum scan angle

x = space length of travel of field from the last aperture element relative to first element to form wavefront at maximum scan angle θ.

τ = equivalent signal pulse length

t = time for signal to travel through distance x

θ_0 = beamwidth in degrees

f_0 = transmission frequency in hertz

B $= \frac{1}{\tau}$ = signal bandwidth

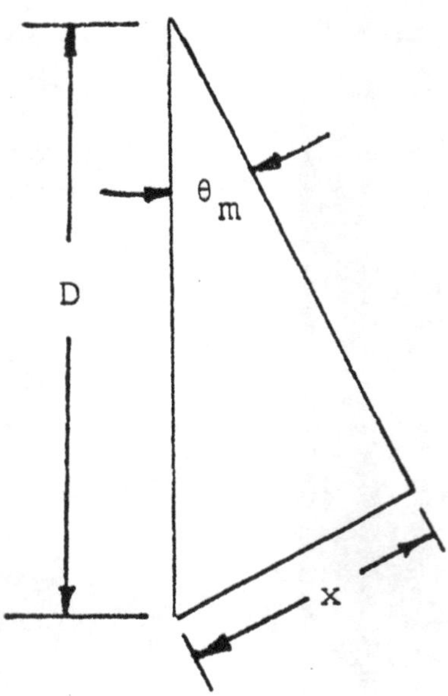

$$x = D \sin \theta_m$$

$$t = \frac{x}{c} = \frac{D \sin \theta_m}{c}$$

Since

$$c = f_0 \lambda$$

and

$$\theta_0 \approx \frac{64\lambda}{D}$$

$$t = \frac{64 \sin \theta_m}{\theta_0 f_0}$$

Practical design is to allow ½ pulse width overlap between transmissions from first and last radiating elements.

$$\therefore t \leqslant \frac{\tau}{2}$$

or

$$\tau \geqslant \frac{128 \sin \theta_m}{\theta_0 f_0}$$

or

$$B \leqslant \frac{\theta_0 f_0}{128 \sin \theta_m}$$

Appendix B

Build-Up Time in a Frequency Scan Array

D = length of array aperture

θ_1 = maximum scan angle in one direction

θ_2 = maximum scan angle in opposite direction

x = space length of travel of field from the last aperture element relative to first element to form wavefront at maximum scan angles.

L = helical or serpentine length

f_1 = lowest frequency used for scanning

f_2 = highest frequency used for scanning

Δf = $f_2 - f_1$ bandwidth required for frequency scanning

τ = equivalent signal pulse length

x' = $x + L$

t = time to travel through distance x

θ_0 = beamwidth in degrees

B = $\frac{1}{\tau}$ = signal bandwidth

c = $f_0 \lambda$

Phase Control for Scanning

$$\theta_1 + \theta_2 = \frac{L}{\Delta\lambda} = \frac{\Delta f L}{c} = \frac{D \sin \theta_1}{\lambda_1} + \frac{D \sin \theta_2}{\lambda_2}$$

or

$$L = \frac{cD}{\Delta f} \left(\frac{\sin \theta_1}{\lambda_1} + \frac{\sin \theta_2}{\lambda_2} \right) = \frac{D}{\Delta f} (f_1 \sin \theta_1 + f_2 \sin \theta_2)$$

Total Length of Travel

$$x' = L + x = L + D \sin\theta_1 = \frac{D}{\Delta f}(f_1 \sin\theta_1 + \Delta f \sin\theta_1 + f_2 \sin\theta_2)$$

$$x' = \frac{D}{\Delta f}\left[\underbrace{(f_1 + \Delta f)}_{f_2}\sin\theta_1 + f_2 \sin\theta_2\right] = \frac{Df_2}{\Delta f}(\sin\theta_1 + \sin\theta_2)$$

since

$$D = \frac{64\lambda}{\theta_0} \quad \text{and } t = \frac{x'}{c}, \; \lambda = \lambda_2$$

$$t = \frac{64}{\theta_0 \Delta f}(\sin\theta_1 + \sin\theta_2)$$

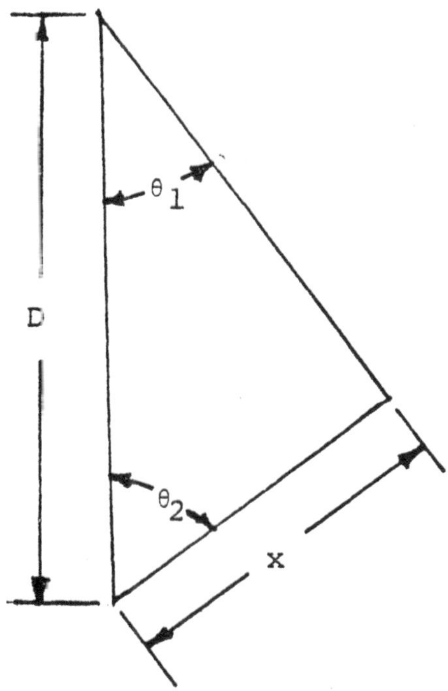

For

$$t \leqslant \frac{\tau}{2}$$

$$\tau \geqslant \frac{128}{\theta_0 \Delta f}(\sin\theta_1 + \sin\theta_2)$$

or

$$B \leqslant \frac{\theta_0 \Delta f}{128(\sin\theta_1 + \sin\theta_2)} = \frac{\theta_0 \Delta f}{256 \sin\theta_m} \quad \text{for } \theta_m = \theta_1 = \theta_2$$

List of Symbols

A	= azimuth
ATT	= attenuator
B	$= \frac{1}{\tau}$ = signal bandwidth
D	= length of array aperture
DIV	= divider
E	= elevation
L	= helical or serpentine length
VMPS	= variable microwave phase shifter
a	= broad dimension of waveguide
c	$= f_o \lambda$
d	= interelement spacing
f	$= \frac{\Delta\gamma}{\Delta t}$ = scan rate
f_o	= transmission frequency in hertz
f_1	= lowest frequency used for scanning
f_2	= highest frequency used for scanning
f_A	= azimuth scanning frequency

f_E	= elevation scanning frequency
f_{IF}	= intermediate frequency
n	$= \frac{s}{\lambda_{go}}$, an integer
s	= path length between radiation slots
t	= time for signal to travel through distance x
x	= space length of travel of field from the last aperture element relative to first element to form wavefront at maximum scan angle θ
x	= x + L
Δ	= difference signal — monopulse
Δ_A	= azimuth difference signal — monopulse
Δ_E	= elevation difference signal — monopulse
$\Delta_{A\,1,2,3\ldots}$	= azimuth difference signal of each horizontal layer
Δ_{A_L}	= difference of azimuth signal on lower half
Δ_{A_U}	= difference of azimuth signal on upper half
Δf	$= f_2 - f_1$ bandwidth required for frequency scanning
$\Delta\phi$	= interelement phase shift
Σ	= sum signal — monopulse
$\Sigma_{1,2,3\ldots}$	= sum signal of each horizontal layer
Σ_{A_L}	= sum of azimuth signal on lower half
Σ_{A_U}	= sum of azimuth signal on upper half
ϕ	= phase of phase shifter
ϕ_E	= differential phase shift for elevation
ϕ_s	= electrical degrees
α	= phase differential between the target return signals (difference is caused by the lengths of the paths from the target to the antenna)
γ	= differential phase shift $= \phi_1 - \phi_2 = \phi_2 - \phi_3$ = space angle
θ	= beam position relative to broadside
θ_o	= beamwidth in degrees
θ_1	= maximum scan angle in one direction
θ_2	= maximum scan angle in opposite direction

θ_m = maximum scan angle

λ = free space wavelength

λ_g = waveguide wavelength corresponding to λ

$$= \frac{\lambda}{\left[1 - \left(\frac{\lambda}{2a}\right)^2\right]^{1/2}}$$

λ_{go} = broadside waveguide wavelength

τ = equivalent signal pulse length

$*$ = conjugate

Detection Theory

Chapter Four

4.1 Statistical Decision Process — The radar sensor has to detect a signal return from a fluctuating target in the presence of noise. A complex target may be considered as a large number of independent reflectors that scatter energy in all directions. The energy scattered in the direction of the radar is of prime interest. The relative phases and amplitudes of the echo signals from the individual scattering objects as measured at the radar receiver determine the total cross section. The behavior varies between total reinforcement and total cancellation. In most radar applications the separation between the individual scatterers is large compared with the wavelength and the phases of the individual signals at the radar receiver will vary as the viewing aspect is changed and cause a scintillating echo.

The net result is a signal return that is fluctuating and competing with noise. At the receiver output a decision has to be made on whether the detected output is due to noise alone or to signal-plus-noise. The decision is made automatically or by a human operator based on the information presented at the radar display console.

Since both noise and target cross section fluctuations are random phenomena, target detection is a statistical decision process.

There are two types of error that the radar can make in each space increment interrogated:

1. It can indicate a target is present when no target exists.

   Noise is mistaken for signal when only noise is present.

This occurs whenever the noise is sufficiently large to exceed the threshold level.

This occurs with relative frequency P_N, where P_N is known as the probability of false alarm.

2. It can indicate that no target is present when a target does exist.

   This is a missed detection and occurs whenever the sum of signal plus noise voltage is not large enough to exceed the threshold level.

   This error occurs with relative frequency $(1-P_D)$ where P_D is the probability of detection.

The system architect makes use of the statistical decision process for design purposes by selecting the optimum processing for detection. In addition, he identifies the target models (if not specified) and designs radar operation to optimize performance against these target models. The signal-to-noise ratio, S/N, to be used in the range equation is selected based on the false alarm probability, P_N, and detection probability, P_D. The cumulative probability of detection and/or amount of integration is optimized (required signal-to-noise ratio is minimized) for the selected target models and target dynamics by varying beam stepping increments, scan rates and instrumented range.

To provide the material required for performing these tasks, several points of interest to the systems engineer are reviewed of the detection process and target models.

4.2 Likelihood Ratio — The threshold setting in a radar receiver is a compromise between the two types of errors made in the decision process. A relatively large threshold will reduce the false alarm rate to a very small value, but there will be more missed detections unless the radar is designed to provide large signal-to-noise ratios on target returns within the radar's coverage volume. In modern automatically operated radars, especially ESRS s, relatively high thresholds are employed to avoid saturation of the radar programming and data processing from excessive response to false alarms. The relative importance of these two types of errors and, therefore, the threshold setting is influenced to a large extent by the radar parameters and radar application.

Detection criteria may be described in terms of the likelihood ratio. It is the ratio of two likelihood functions.

$$L(y) = \frac{P_s(y)}{P_o(y)} \qquad (4.1)$$

where

$L(y)$	=	likelihood ratio
y	=	$\begin{cases} S + N & \text{signal-plus-noise} \\ N & \text{noise alone} \end{cases}$ continuous random variable
$P_s(y)$	=	probability of y given s conditional probability
$P_o(y)$	=	probability of y with no signal present conditional probability

The likelihood ratio may also be defined as the ratio of the probability-density function of signal-plus-noise, $p_s(y)$, to the probability density function of noise alone $p_o(y)$. It is an important statistical tool because it is a measure of how likely it is that the envelope of the continuous random variable, y, in the receiver is due to signal-plus-noise as compared to noise alone. Several detection criteria are considered.

1. *Bayes decision rule* minimizes the average loss defined as

$$L_A = q\alpha C_\alpha + p\beta C_\beta \tag{4.2}$$

where

L_A	=	average loss
q	=	a priori probability that there is no signal
p	=	a priori probability that signal is present
$p + q$	=	1
α	=	probability of false alarm, P_N
β	=	probability that signal is absent, $(1-P_D)$
C_α	=	cost of Type 1 error (positive quantity)
C_β	=	cost of Type 2 error (positive quantity)

and obtains

$$L(y) = \frac{P_s(y)}{P_o(y)} \geqslant \frac{C_\alpha}{C_\beta} \quad \text{signal is present.} \tag{4.3}$$

2. *The ideal observer decision rule* minimizes the total probability of error.

$$P_o = p\beta + q\alpha \tag{4.4}$$

This solution corresponds to a Bayes decision in which the cost of errors are assigned equal weights ($C_\alpha = C_\beta$).

Therefore:

$$L(y) = \frac{P_s(y)}{P_o(y)} \geqslant 1 \quad \text{signal is present.} \tag{4.5}$$

3. The Neyman-Pearson criterion does not require knowledge of a priori signal statistics, nor explicit assignment of cost functions. The Neyman-Pearson criterion sets a fixed false-alarm probability, α, and minimizes the missed target detection probability, β, (maximizes probability of detection, P_D).

$$L(y) = \frac{P_s(y)}{P_o(y)} \geqslant K \quad \text{signal is present.} \tag{4.6}$$

where K is a real positive number dependent on the specification of false-alarm probability α.

4. The minimax criterion does not require knowledge of a priori signal statistics. The minimax decision rule corresponds to a Bayes rule for a priori signal statistics that maximize Bayes loss. This approach is very conservative since the worst a priori state of nature is assumed.

$$L(y) = \frac{P_s(y)}{P_o(y)} \geqslant \frac{C_\alpha}{C_\beta} \quad \text{signal is present.} \tag{4.7}$$

where L(y) is obtained for a priori signal statistics that maximize Bayes loss.

All the preceding decision rules are based on a comparison of the likelihood ratio, L(y), with different threshold values. A receiver that computes L(y) for all possible values of the continuous random variable y and compares this value with an adjustable threshold can perform any of the preceding tests.

It can be shown that the structure of this receiver in presence of white gaussian noise consists of a matched filter followed by an envelope detector whose output is compared with a threshold to arrive at a decision.

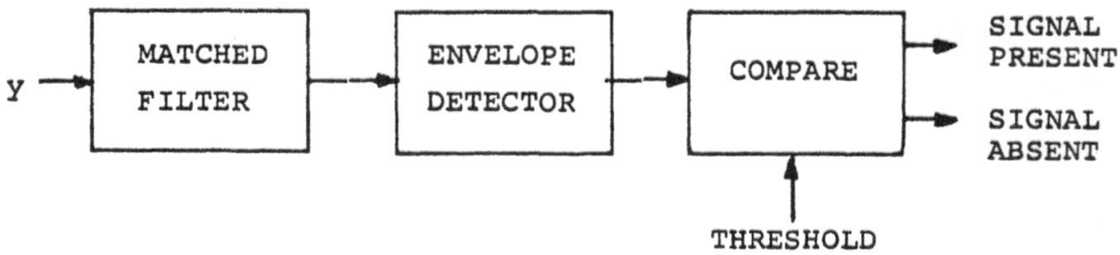

4.3 Target Models and Applications — The signal-to-noise power ratio to be used in the range equation is determined from the false alarm probability P_N and detection probability P_D for a particular target fluctuation model. The probability density function must be known to account for target cross-section fluctuations. The correlation properties of the target with time are required to determine the degree of correlation of the cross section with time or number of pulses. Complete data of this type is too complex to obtain for general radar applications. Instead general statistical fluctuation models have been developed and used for radar design. The constant amplitude model have been developed and used for radar design. The constant amplitude model is for a nonfluctuating cross section. Swerling has developed four target fluctuation models that are commonly used. In two of the models it is assumed that the fluctuations are completely correlated pulse-to-pulse during a particular scan but are completely uncorrelated from scan-to-scan. In the other two models the fluctuations are assumed to be very rapid and completely uncorrelated pulse-to-pulse. Other cases are also discussed.

Target Models — The following target models are used in radar design:

1. Constant Amplitude (Case O)
2. SWI — Rayleigh
   — Constant — Pulse-to-Pulse
   — Incoherent Fluctuation — Scan-to-Scan
3. SWII — Rayleigh
   — Incoherent Fluctuation — Pulse-to-Pulse
4. SWIII — One Dominant + Rayleigh
   — Constant — Pulse-to-Pulse
   — Incoherent Fluctuation — Scan-to-Scan
5. SWIV — One Dominant + Rayleigh
   — Incoherent Fluctuation — Pulse-to-Pulse
6. Log-Normal
7. Weibull

The likelihood ratio indicates that the envelope of the signal determines detection. Therefore, the probability density functions (pdfs) are given for envelope distributions. Square law detection is employed for all the target models.

1. Constant Amplitude

$$p(R)_c = \frac{R}{N} e^{-\left[\frac{R^2 + S^2}{2N}\right]} I_o\left(\frac{SR}{N}\right) \tag{4.8}$$

where:

$p(R)_c$ = Rician pdf for envelope distribution of narrow band modulated sinusoidal signal of constant amplitude in the presence of noise for single hit

R = signal + noise voltage

S = signal voltage

N = average noise power (mean-square noise voltage)

$I_o\left(\frac{SR}{N}\right)$ = modified Bessel function, which results from averaging (integrating) out the random phase variation

For single hit,

$$P_{OD} = \int_{y_b}^{\infty} p(R)_c \, dR \tag{4.9}$$

and

$$P_{ON} = \int_{y_b}^{\infty} p(R)_c \Big|_{S=0} dR \tag{4.10}$$

where:

P_{OD} = single hit probability of detection for constant amplitude target model — Case O

P_{ON} = single hit false alarm rate for constant amplitude target model — Case O

$$y_b = \frac{T}{\sqrt{2N}} \tag{4.11}$$

and

T = threshold voltage level

For n hit integration (square law post detection integration),

$$C_{On}(-p) = [C_C(-p)]^n \quad \text{characteristic function} \tag{4.12}$$

and

$$p_{On}(R)_c = \int_{-\infty}^{\infty} C_{On}(-p)\, e^{pR}\, dp \tag{4.13}$$

Inverse Fourier Transform

$$\therefore P_{OD_n} = \int_{y_b}^{\infty} p_{On}(R)_c\, dR \tag{4.14}$$

and

$$P_{ON_n} = \int_{y_b}^{\infty} P_{on}(R)_c \Big|_{S=0} dR \tag{4.15}$$

where

$$C_{OC}(-p) = \int_{0}^{\infty} p(R)_c\, e^{-pR}\, dR \quad \text{Fourier Transform} \tag{4.16}$$

$p_{On}(R)_c$ = pdf of constant amplitude target model in presence of noise for n hits — Case O.

P_{OD_n} = probability of detection of constant amplitude target model for n hits — Case O.

P_{ON_n} = false alarm rate of constant amplitude target model for n hits — Case O.

2. SWI — Rayleigh

For fluctuating target model

$$p(R)_F = \int p(\sigma)\, p(R)_c\, d\sigma$$

where target fluctuates according to independent samples from pdf $p(\sigma)$

$p(R)_F$ = pdf for fluctuating target model in presence of noise for single hit

$$p_1(\sigma) = \frac{1}{\sigma_o} e^{-\frac{\sigma}{\sigma_o}} \tag{4.17}$$

= pdf for Rayleigh target cross section model

σ_o = average cross section

For single hit

$$P_{1D} = \int_{y_b}^{\infty} p_1(R)_F \, dR \tag{4.18}$$

$$P_{1N} = \int_{y_b}^{\infty} p_1(R)_F \Big|_{S=0} dR \tag{4.19}$$

where

$p_1(R)_F$ = pdf for Rayleigh target cross section model in presence of noise for single hit — Case 1

P_{1D} = single hit probability of detection for SWI — Rayleigh target model — Case 1

P_{1N} = single hit false alarm rate for SWI — Rayleigh target model — Case 1

For n hit integration (square law post detection integration),

$$C_{1n}(-p) = \overline{[C_{1F}(-p)]^n} = \int_o^{\infty} [C_{1F}(-p)]^n \, p_1(\sigma) \, d\sigma \tag{4.20}$$

and

$$p_{1n}(R)_F = \int_{-\infty}^{\infty} C_{1n}(-p) \, e^{pR} \, dp \tag{4.21}$$

$$\therefore\ P_{1D_n} = \int_{y_b}^{\infty} p_{1n}(R)_F \, dR \tag{4.22}$$

$$P_{1N_n} = \int_{y_b}^{\infty} p_{1n}(R)_F \Big|_{S=0} dR \tag{4.23}$$

where

$$C_{1F}(-p) = \int_{o}^{\infty} p_1(R)_F \, e^{-pR} \, dR \tag{4.24}$$

$p_{1n}(R)_F$ = pdf for SWI — Rayleigh target model in presence of noise for n hits — Case 1

P_{1D_n} = probability of detection for SWI — Rayleigh target model for n hits — Case 1

P_{1N_n} = false alarm rate for SWI — Rayleigh target model for n hits — Case 1

3. SWII — Rayleigh

For single hit — the same formulas and results apply as for SWI — Rayleigh. The two cases are identical for single hit.

$$p_2(\sigma) = p_1(\sigma) = \frac{1}{\sigma_o} e^{-\frac{\sigma}{\sigma_o}} \tag{4.25}$$

$$p_2(R)_F = p_1(R)_F \tag{4.26}$$

$$P_{2D} = P_{1D} \tag{4.27}$$

$$P_{2N} = P_{1N} \tag{4.28}$$

For n hit integration (square law post detection integration),

$$C_{2n}(-p) = \overline{[C_{2F}(-p)]^n} = \left[\int_{o}^{\infty} C_{2F}(-p)\, p_2(\sigma) d\sigma\right]^n \tag{4.29}$$

or

$$C_{2n}(-p) = \left[C_{1n}(-p) \Big|_{n=1} \right]^n \tag{4.30}$$

and

$$p_{2n}(R)_F = \int_{-\infty}^{\infty} C_{2n}(-p)\, e^{pR}\, dp \tag{4.31}$$

$$\therefore P_{2D_n} \int_{y_b}^{\infty} p_{2n}(R)_F\, dR \tag{4.32}$$

$$P_{2N_n} = \int_{y_b}^{\infty} p_{2n}(R)_F \Big|_{S=0} dR \tag{4.33}$$

where

$$C_{2F}(-p) = C_{1F}(-p) = \int_{o}^{\infty} p_1(R_F)\, e^{-pR}\, dR \tag{4.34}$$

$p_{2n}(R)_F$ = pdf for SWII — Rayleigh target model in presence of noise for n hits — Case 2

P_{2D_n} = probability of detection for SWII — Rayleigh target model for n hits — Case 2

P_{2N_n} = false alarm rate for SWII — Rayleigh target model for n hits — Case 2

4. SWIII — One Dominant + Rayleigh

The mathematical procedure is identical to that of SWI — Rayleigh. The only difference is the use of the One Dominant + Rayleigh pdf for the target cross section model.

$$p_3(\sigma) = \frac{4\sigma}{(\sigma_o)^2}\, e^{-\frac{2\sigma}{\sigma_o}} \tag{4.35}$$

where

$p_{,}(\sigma)$ = pdf for One Dominant + Rayleigh target cross section model

σ_o = average cross section

5. SWIV — One Dominant + Rayleigh

The mathematical procedure is identical to that of SWII — Rayleigh. The only difference is the use of the One Dominant + Rayleigh pdf for the target cross section model.

$$p_4(\sigma) = p_3(\sigma) = \frac{4\sigma}{(\sigma_o)^2} e^{-\frac{2\sigma}{\sigma_o}} \tag{4.36}$$

6. Log-Normal

$$p_5(\sigma) = \frac{1}{\sqrt{2\pi}\,\sigma_s\,\sigma} \exp -\left[\frac{(\ln\sigma - \ln\sigma_m)^2}{2\sigma_s^{\,2}}\right] \tag{4.37}$$

where

$p_5(\sigma)$ = pdf for log-normal target cross section model

σ_s = standard deviation of ln σ

σ_m = median cross section

$$\sigma_o = \sigma_m \exp\left[-\frac{\sigma_s^{\,2}}{2}\right] = \text{average cross section} \tag{4.38}$$

7. Weibull

$$p_6(\sigma) = \frac{1}{a\,b\,\sigma^{(1-\frac{1}{a})}} \exp\left[-\left(\frac{\sigma}{b}\right)^{1/a}\right] \tag{4.39}$$

where

$p_6(\sigma)$	= pdf for Weibull target cross section model	
σ_s	= standard deviation of σ	
a	= spread parameter — σ_s and a describe spread of distribution. The Weibull a = 1 model is identical to SWI, II — Rayleigh.	
σ_m	= median cross section	
b	$= \left(\dfrac{\sigma_m}{\ln 2}\right)^{\frac{1}{a}}$ = slope parameter	(4.40)
σ_o	$= b^a\Gamma(1+a)$ = average cross section where Γ denotes the gamma function.	(4.41)

The mathematical procedure to structure the integrals for determining P_D as a function of P_N, S/N and n for the various target models and the two cases of slow and fast time varying target cross section has been developed to familiarize the reader with the statistical decision theory approach to radar design and the inherent assumptions and limitations involved.

The pdfs have been formulated for square law envelope detection. The optimum ln I_o detector law is approximated by a square law envelope detector at small S/N and by a linear detector when the single-pulse S/N is large. The difference between the square law and the linear detectors was shown by Marcum[1] to be negligible for a typical constant amplitude target model for $P_D = 0.50$ and $(P_N/n) = 10^{-6}$. For small n ($n \leqslant 70$) the linear detector is superior by an amount not exceeding 0.11 dB. As $n \rightarrow \infty$ the square law detector is superior by an amount assymplotically approaching 0.19 dB. In practical implementation the envelope detector exhibits a square law characteristic for small signals and a linear characteristic for large signals. The performance difference between the two detector characteristics is usually of little concern to the radar system designer.

The two cases considered for the fluctuation models assume either complete correlation between the pulses in any particular scan or either complete independence between the pulses. On one extreme the target orientation is assumed to change sufficiently from scan-to-scan so that echo amplitude and initial phase are statistically independent in each scan. In the other extreme, the target orientation is assumed to change rapidly in an interpulse period (pulse-to-pulse) but slow when compared to the duration of a single pulse which results in a statistically independent amplitude and initial phase for each pulse in the train but practically no distortion of the waveform structure within a pulse. The more likely situation is that the pulses of a particular scan will lie somewhere within these two extremes and be partially correlated. Fluctuations that lie between these extremes are difficult to analyze and only limited results have been obtained.

Marcum[1] solved the integrals for determining P_D as a function of P_N, S/N, and n for a constant amplitude target model in the presence of noise and graphed the results. Swerling[2] extended this work to include his four fluctuating target models. DiFranco and Rubin[3] computed and graphed these 5 cases for additional values of P_N and n and in more convenient form for use by the radar designer. Meyer and Mayer[4] provides even more extensive tabulation for all 5 cases.

Therefore, this application of statistical decision theory to radar design is reduced to simply selecting P_N and n for a desired or specified P_D and target model; then determing the single-hit S/N, as obtained from any of the graphed or tabulated data sources mentioned above, for use in the radar range equation.

Applications — The target models discussed are felt to be typical of situations likely to be met in practice, or at least, to bracket a wide range of practical cases.

The amplitude distribution applies when the target aspect angle is unknown. Initial phase is treated as a random variable. It is assumed to have a (least favorable) uniform distribution from 0 to 2π.

The amplitude distribution also applies when the target is too irregular to express radar cross section as a geometric function. Instead, the radar cross section is expressed statistically. The average value of the radar cross section is used in the radar range equation.

The normal procedure is to select the target model for radar design that most closely approximates the actual target fluctuations. In some applications several target models are used to cover detection of aircraft, missiles and satellites. In other applications it might be appropriate to consider the specified targets to be intermediate between two theoretical models.

1. Constant amplitude — This model applies for spherical or nearly spherical objects and point targets. It characterizes stationary and moving targets that display an unchanging aspect (radar cross section) on successive pulse transmissions.

1. Satellites (spherical)
2. Balloons
3. Meteors

2. SW I, II — Rayleigh — This model applies to a target consisting of many independently fluctuating scatterers of approximately equal echo area. In theory the number of independent scatterers is infinite, in practice this model applies adequately for any target composed of four or more scatterers. Objects which are large compared to wavelength (and are not shaped too much like a sphere) conform to this model.

1. Aircraft — large surface (jet, propeller and helicopter)
2. Rain clutter
3. Surface clutter (viewed with grazing angle greater than 5 degrees)

3. SW III, IV — One Dominant + Rayleigh — This model applies to a target represented as one large scatterer with one other small scatterers or one large scatterer subject to small changes in orientation.

1. Missiles, 2. Aircraft, 3. Rockets } long narrow surfaces
4. Spherical satellites with extensions

4. SW I, III — Constant — (pulse-to-pulse)
Incoherent Fluctuations — (scan-to-scan)

These models apply to jet aircraft or missiles and radars having high pulse repetition rates and high scan rates.

5. SW II, IV — Incoherent Fluctuations — (pulse-to-pulse) — These models apply to propeller aircraft and helicopters if the propellers contribute a large portion of the echoing area. They also apply to targets for which very small changes in orientation cause large changes in echoing area, such as long thin objects at high frequency or targets viewed by radars with sufficiently low repetition rates.

These models always apply to complex targets when radars employ pulse-to-pulse frequency agility (different transmit frequency on each pulse).

6. Weibull and Log-Normal — These models apply to targets with a broad spectrum of amplitudes.

1. Surface clutter (viewed at low grazing angles by a medium or high resolution radar)
2. Missile tanks (cylindrical)

There are many targets that are intermediate between these various cases.

4.4 Comparison of Target Models —

Single Hit — The particular target fluctuation model employed to design a radar system to meet a specified false alarm rate, P_N, and detection probability, P_D, on a single hit basis determines the signal-to-noise ratio, S/N, which in turn directly influences the sizing of the radar.

Table 4.1 contains the single hit S/N required as a function of P_D for 10^{-3}, 10^{-6} and 10^{-9} values of P_N for the constant amplitude and Swerling target models. This data is plotted in Figs. 4.1, 4.2 and 4.3 to indicate the fluctuation loss/gain relative to a constant amplitude model for the various Swerling cases as functions of P_N and P_D. Several points of interest may be obtained from this data useful to radar designers:

1. The fluctuation loss is negligible at $P_D \approx 0.34$ for all the Swerling target models and for all values of P_N.
2. For $P_D < 0.34$ there is a fluctuation gain and for $P_D >$ 0.34 there is a fluctuation loss for all the Swerling target models and for all values of P_N.
3. The fluctuation loss increases as P_D increases above 0.34 for all the Swerling target models and for all values of P_N.
4. The fluctuation gain increases as P_D decreases below 0.34 for all the Swerling target models and for all values of P_N.
5. The fluctuation gain/loss increases as P_N decreases for a specified P_D (not equal to 0.34).
6. The fluctuation gain/loss is more for the Swerling I, II target models than for the Swerling III, IV target models for any specified values of P_D (not equal to 0.34) and P_N.

Another important area affected by single hit target fluctuation model characteristics is low sampling rate (approximately once per second) target tracking. The multitarget tracking capability of phased arrays is dependent on relatively low sampling rates. The lower the sampling rate on individual targets the larger the number of targets that can be tracked simultaneously on a time-shared basis. A system designed to have, in the track mode, power and time budgets allowing 100 looks per second can track 100 targets at one look per second or only 10 targets at 10 looks per second. Consequently, the sampling rate on individual target tracks is made as low as possible consistent with maintaining "tenacity of track" for the specified target dynamics. Track maintenance is not only affected by target dynamics but it is affected also by the fading characteristics of the target. Table 4.2 and Fig. 4.4 contain the cumulative probability versus amplitude distribution relative to the median for the Swerling I and III target fluctuation models.

It is noted in Fig. 4.4 that a Swerling I target model has a higher probability of fading than a Swerling III target model. For example, a target whose fluctuation characteristics are represented by a Swerling I target model has a 6.7% probability of fading 10 dB relative to the median value whereas a Swerling III type target has only a 1.3% probability of fading the same amount. It follows that the target models must be realistically selected and accounted for in the design of low sampling rate multitarget tracking loops.

Table 4-1

S/N vs. P_N, P_D

P_N	P_D	Constant S/N dB	SW I, II S/N dB	Δ dB	SW III, IV S/N dB	Δ dB
10^{-3}	.05	2.4	1.1	-1.3	1.6	-0.8
	.10	4.1	3.0	-1.1	3.4	-0.7
	.20	5.7	5.2	-0.1	5.3	-0.4
	.34	7.0	7.3	.3	7.0	0
	.50	8.1	9.5	1.4	8.7	0.6
	.80	9.9	14.8	4.9	12.3	2.4
	.90	10.8	18.1	7.3	14.4	3.6
	.99	12.5	28.4	15.9	20.2	7.7
10^{-6}	.05	7.8	5.4	-2.4	6.4	-1.4
	.10	8.7	6.8	-1.9	7.6	-1.1
	.20	9.7	8.7	-1.0	9.1	-0.6
	.34	10.5	10.8	0.3	10.5	0
	.50	11.2	12.8	1.6	12.0	0.8
	.80	12.6	17.9	5.3	15.3	2.7
	.90	13.2	21.1	7.9	17.3	4.1
	.99	14.5	31.4	16.9	23.0	8.5
10^{-9}	.05	10.4	7.7	-2.7	8.5	-1.9
	.10	11.1	9.0	-2.1	9.6	-1.5
	.20	11.8	10.8	-1.0	11.0	-0.8
	.34	12.4	12.5	0.1	12.4	0
	.50	13.1	14.6	1.5	13.8	0.7
	.80	14.1	19.6	5.5	17.1	3.0
	.90	14.7	22.9	8.2	19.0	4.3
	.99	15.8	33.2	17.4	24.6	8.8

Table 4-2

Cumulative Probability Versus Amplitude Distribution

σ/σ_m dB	σ/σ_m	P_{D_1}	P_{D_3}
-20	.01	.007	.0001
-18	.0158	.011	.0003
-16	.0251	.017	.0009
-14	.0398	.027	.0021
-12	.0631	.043	.0052
-10	.1000	.067	.0126
- 8	.1585	.104	.0297
- 6	.2512	.160	.0674
- 4	.3981	.241	.1448
- 2	.6310	.354	.2859
0	1	.500	.5000
2	1.5849	.667	.7440
4	2.5119	.825	.9230
6	3.9811	.937	.9903
8	6.3096	.987	
10	10.0000	.999	

SW I $$P_1(\sigma) = 1 - e - \frac{.6931\sigma}{\sigma_m}$$

SW III $$P_3(\sigma) = 1 - \left(1 + \frac{1.6783\sigma}{\sigma_m}\right) e - \frac{1.6783\sigma}{\sigma_m}$$

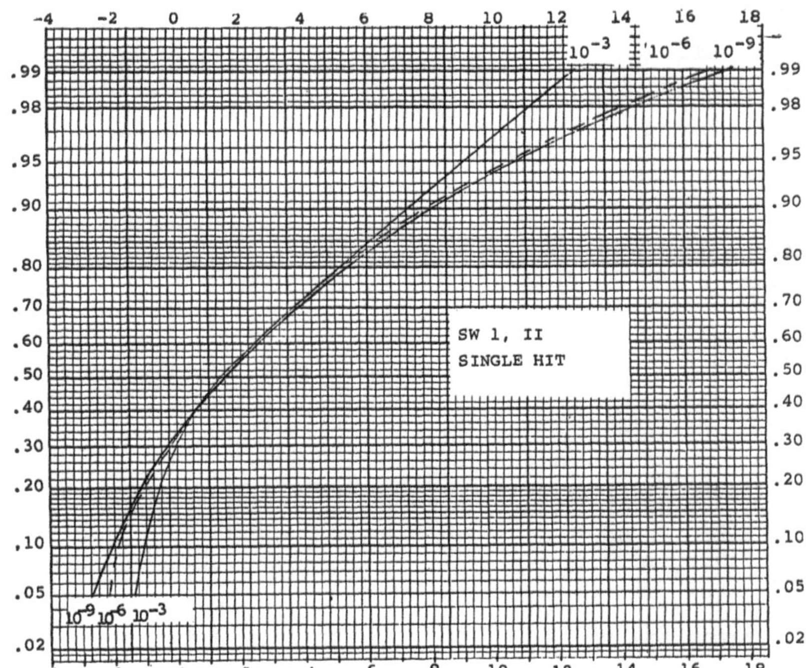

Figure 4-1

Fluctuation Loss — dB

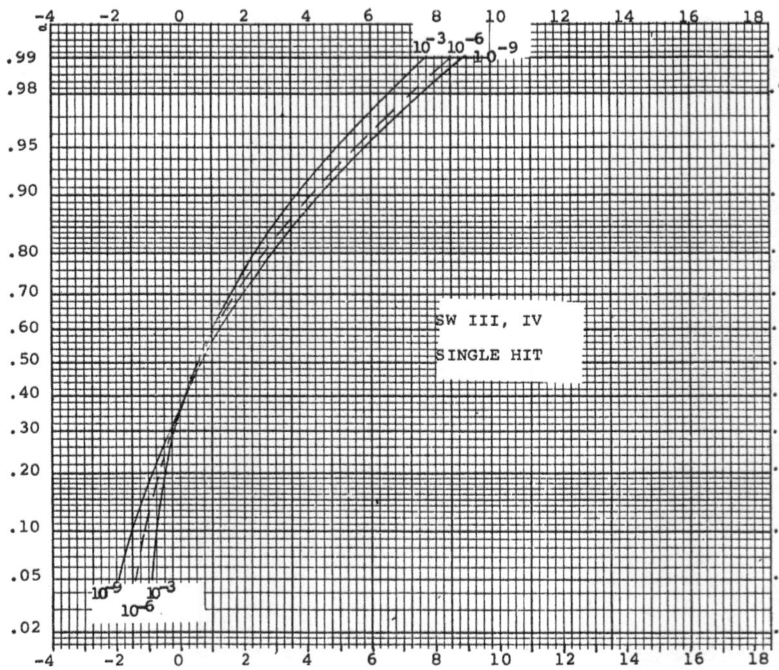

Figure 4-2

Fluctuation Loss — dB

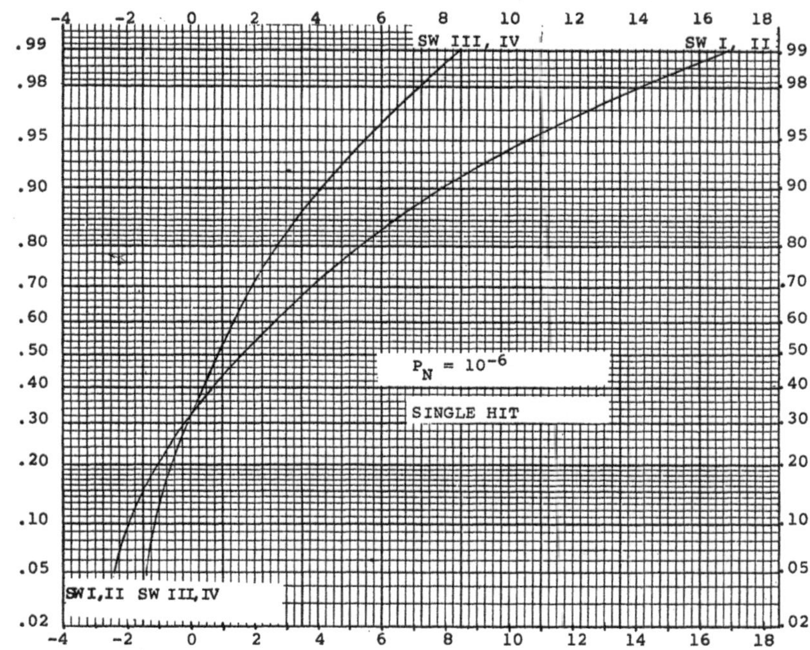

Figure 4-3

Fluctuation Loss — dB

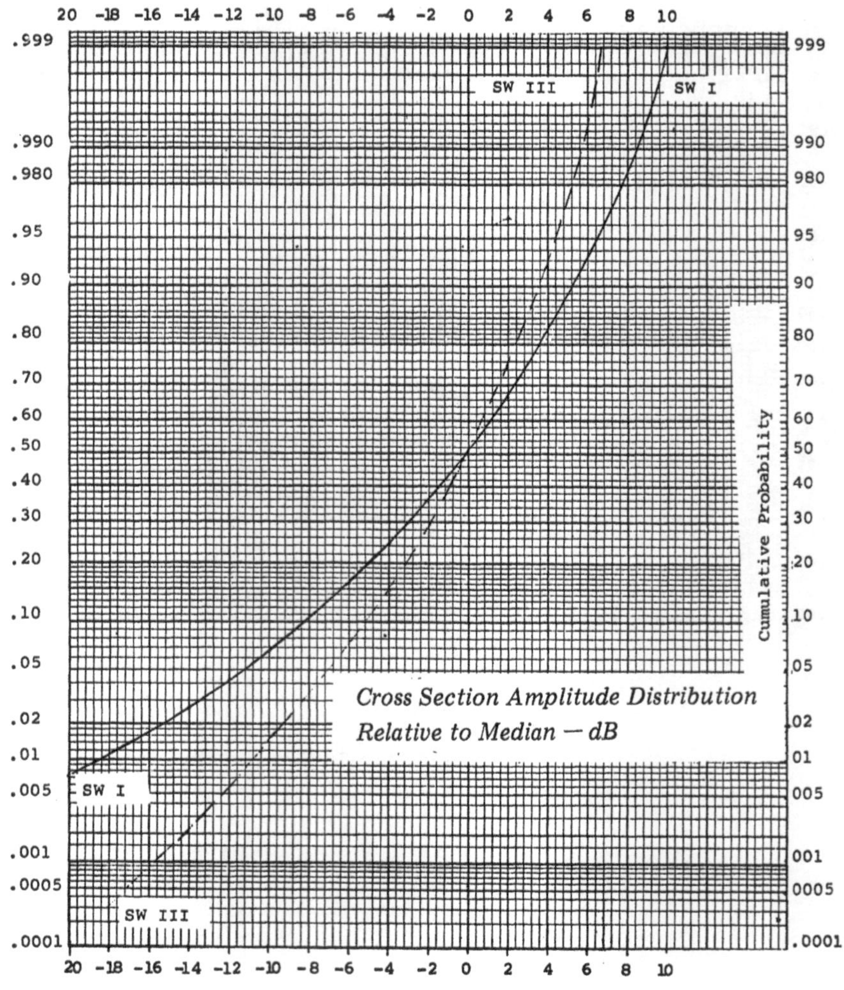

20 -18 -16 -14 -12 -10 -8 -6 -4 -2 0 2 4 6 8 10
.999
.990
.980
.95
.90
.80
.70
.60
.50
.40
.30
.20
.10
.05
.02
.01
.005
.001
.0005
.0001
SW III
SW I
Cumulative Probability
Cross Section Amplitude Distribution
Relative to Median — dB

Figure 4-4

Multihit Integration — In many phased array radar applications the search function is constrained to relatively short frame times (several seconds) to prevent fast moving targets (MACH 2 or more) from passing undetected through the radar coverage sector (typically one quadrant of a hemisphere extending 200 n.miles in range) or to allow adequate reaction time to take prevention measures. In quite a few of these cases there is insufficient time for multihit detection per search beam position. However, in circumstances or designs where adequate search frame time is available, the optimum radar system (in terms of minimum power-aperture) provides for integration of all energy returns during time-on-target. Coherent integration is the most efficient for steady or correlated targets, but post detection (noncoherent) integration is widely used because it is simpler to implement. There are also circumstances where the target dynamics permit longer noncoherent integration than coherent integration during time-on-target. In some cases a combination of coherent and noncoherent integration is employed.

The S/N required for multihit post detection integration to meet specified P_D and P_N is dependent on the number of hits, n, integrated and the target fluctuation model. It is, therefore, of value to the radar designer to become familiar with the characteristics of the various target fluctuation models in terms of S/N, n, P_D and P_N.

Table 4.3 contains data comparing the four Swerling to the constant amplitude target models in terms of S/N versus P_D for $P_N = 10^{-6}$ and n = 10, 100 and 1000. Figs. 4.5, 4.6, and 4.7 are graphical representations of this data. The comparison data represents a combination of integration and fluctuation losses. Table 4.4 contains only the integration loss and was prepared from data contained in Tables 4.1 and 4.3 using the definition of integration loss

$$L_i = \frac{n(S/N)_i}{(S/N)_{SH}} \tag{4.42}$$

or in decibels

$$L_{i_{dB}} = 10 \lg n + 10 \lg (S/N)_i - 10 \lg (S/N)_{SH} \tag{4.43}$$

where:

L_i = integration loss relative to coherent integration

n = number of hits integrated non-coherently

$(S/N)_{SH}$ = single hit S/N required to produce a given P_D and P_N
$(S/N)_i$ = S/N required to produce the same P_D and P_N for n integrated hits

The integration loss provides a measure of decrease in detectability incurred by incoherent processing.

Table 4-3

S/N vs. P_D, n

$P_N = 10^{-6}$

	P_D	Constant S/N dB	SW I S/N dB	Δ dB	SW II S/N dB	Δ dB	SW III S/N dB	Δ dB	SW IV S/N dB	Δ dB
n=10	.05	1.0	-1.6	-2.6	0.4	-0.6	-0.9	-1.9	0.6	-0.4
	.10	1.7	-0.4	-2.1	1.2	-0.5	0.2	-1.5	1.4	-0.3
	.20	2.4	1.4	-1.0	2.2	-0.2	1.5	-0.9	2.3	-0.1
	.34	3.1	2.8	-0.3	3.0	-0.1	3.0	-0.1	3.0	-0.1
	.50	3.7	5.4	1.7	3.8	0.1	4.4	0.7	3.7	0
	.80	4.8	10.4	5.6	5.5	0.7	7.6	2.8	5.2	.4
	.90	5.2	13.6	8.4	6.4	1.2	9.6	4.4	5.8	.6
	.99	6.5	24.0	17.5	8.5	2.0	15.4	8.9	7.5	1.0
n=100	.05	-4.8	-7.8	-3.0	-4.8	0	-7.2	-2.4	-4.8	0
	.10	-4.3	-6.5	-2.2	-4.3	0	-5.8	-1.5	-4.2	0.1
	.20	-3.7	-4.8	-1.1	-3.6	0.1	-4.5	-0.8	-3.6	0.1
	.34	-3.0	-3.0	0	-3.0	0	-3.2	-0.2	-3.0	0
	.50	-2.6	-1.0	1.6	-2.5	0.1	-1.8	0.8	-2.6	0
	.80	-1.7	4.0	5.7	-1.5	0.2	1.4	3.1	-1.5	0.2
	.90	-1.3	7.2	8.5	-1.1	0.2	3.3	4.6	-1.1	0.2
	.99	-0.4	17.5	17.9	-0.1	0.3	9.0	9.4	-0.2	0.2
n=1000	.05	-10.0	-13.2	-3.2			-12.2	-2.2		
	.10	-9.5	-11.8	-2.3			-11.2	-1.7		
	.20	-8.9	-10.2	-1.3			-9.9	-1.0		
	.34	-8.4	-8.5	-0.1			-8.5	-0.1		
	.50	-8.0	-6.5	1.5			-7.2	0.8		
	.80	-7.2	-1.5	5.7			-4.0	3.2		
	.90	-6.8	1.8	8.6			-2.1	4.7		
	.99	-6.1	12.0	18.1			3.5	9.6		

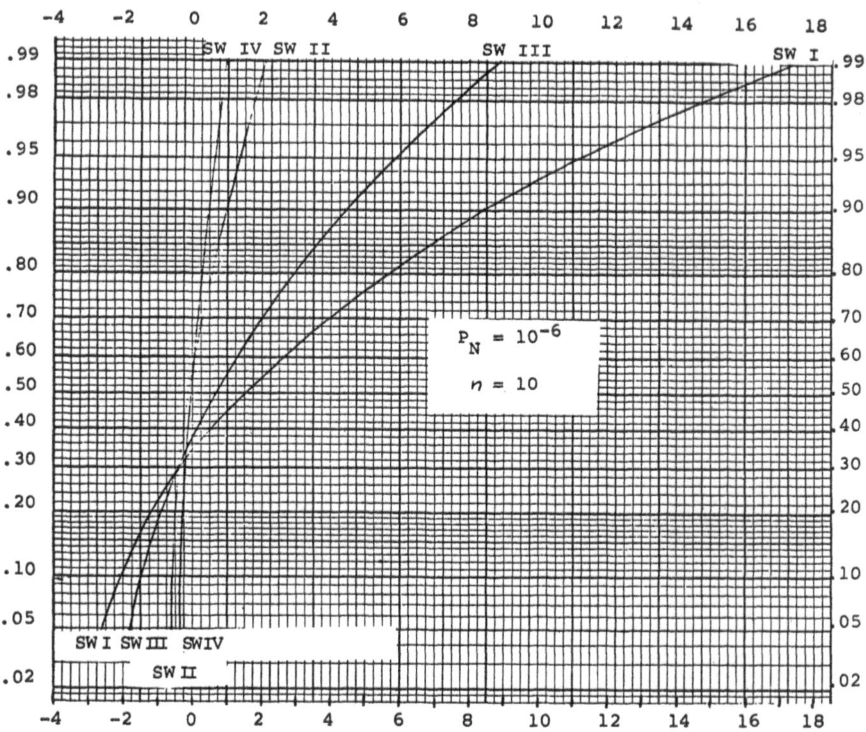

Figure 4-5
Total loss in dB
Integration and Fluctuation Losses
Relative to Constant Amplitude Target Model

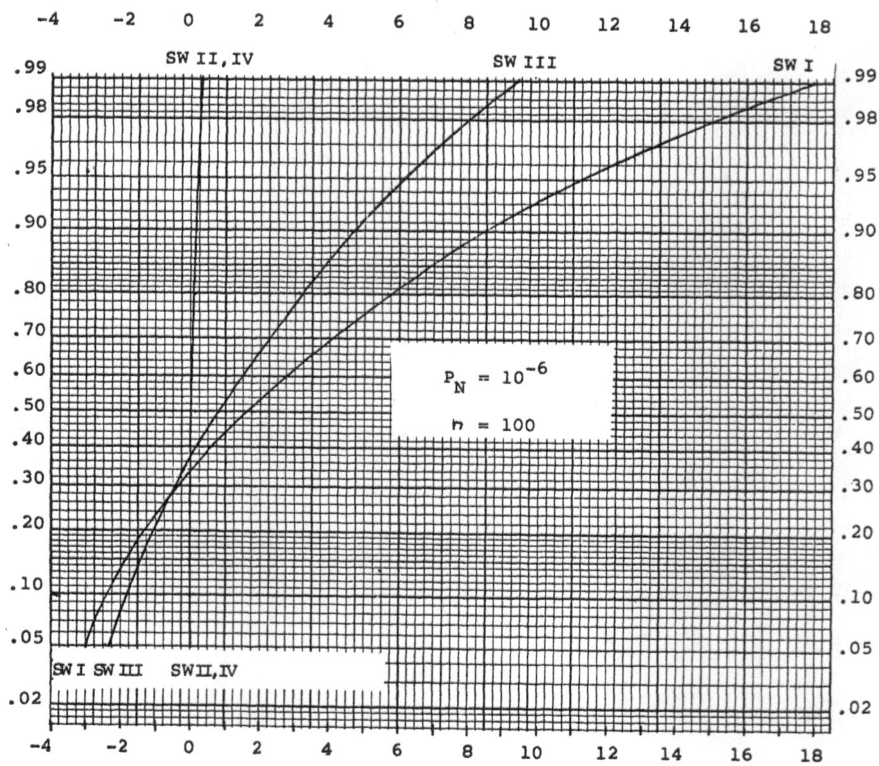

Figure 4-6
Total loss in dB
Integration and Fluctuation Losses
Relative to Constant Amplitude Target Model

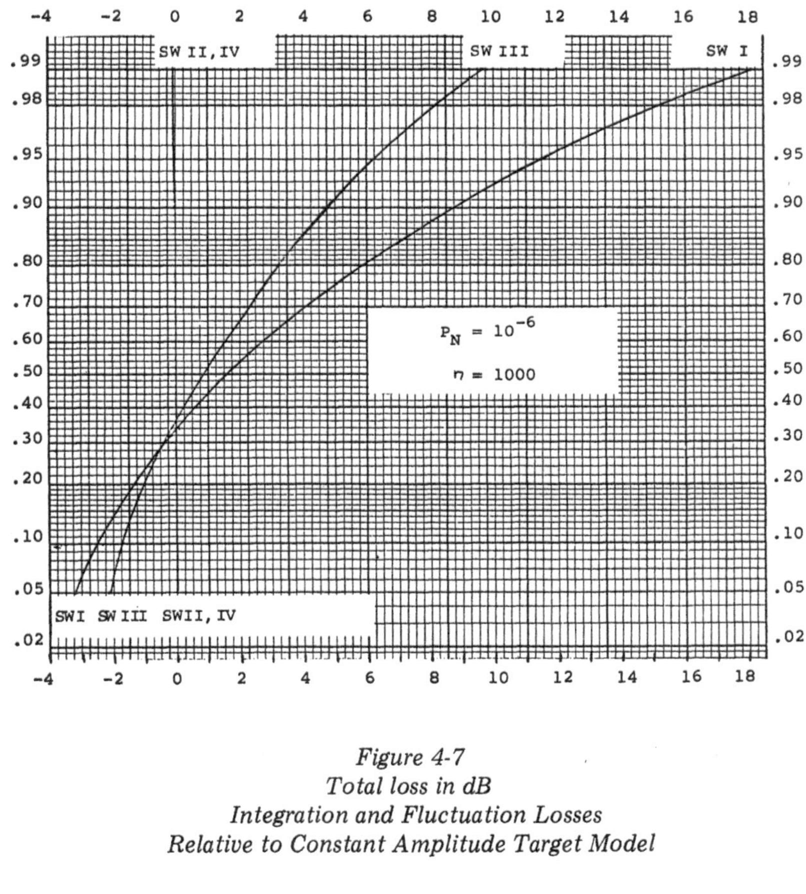

Figure 4-7
Total loss in dB
Integration and Fluctuation Losses
Relative to Constant Amplitude Target Model

Table 4-4
Integration Losses
$P_N = 10^{-6}$

n	P_D	Constant dB	SW I dB	SW II dB	SW III dB	SW IV dB
10	.05	3.2	3.0	5.0	2.7	4.2
	.10	3.0	2.8	4.4	2.6	3.8
	.20	2.7	2.7	3.5	2.6	3.2
	.34	2.6	2.5	2.2	2.5	2.5
	.50	2.5	2.6	1.0	2.4	1.7
	.80	2.2	2.5	- 2.4	2.3	-0.1
	.90	2.0	2.5	- 4.7	2.3	-1.5
	.99	2.0	2.6	-12.9	2.4	-5.5
100	.05	7.4	6.8	9.8	6.4	8.8
	.10	7.0	6.7	8.9	6.6	8.2
	.20	6.6	6.5	7.7	6.4	7.3
	.34	6.5	6.2	7.2	6.3	6.5
	.50	6.2	6.2	4.7	6.2	5.4
	.80	5.7	6.1	0.6	6.1	3.2
	.90	5.5	6.1	- 2.2	6.0	1.6
	.99	5.1	6.1	-11.5	6.0	-3.2
1000	.05	12.2	11.4	14.6	11.4	13.6
	.10	11.8	11.4	13.7	11.2	12.9
	.20	11.4	11.1	12.4	11.0	12.0
	.34	11.1	10.7	10.8	11.0	11.1
	.50	10.8	10.7	9.2	10.8	10.0
	.80	10.2	10.6	4.9	10.7	7.5
	.90	10.0	10.7	2.1	10.6	5.9
	.99	9.4	10.6	- 7.5	10.5	0.9

The following multihit integration characteristics may be obtained from this data useful to radar designers.

1. The combination of fluctuation and integration losses relative to the constant amplitude case is negligible at $P_D \approx 0.34$ for all the Swerling target models and for all values of n (and P_N — not demonstrated).
2. For $P_D < 0.34$ these losses are less than the constant amplitude case and for $P_D > 0.34$ they are more than the constant amplitude case for all the Swerling target models and for all values of n (and P_N).
3. The combination of fluctuation and integration losses is more than and increase relative to the constant amplitude case as P_D increases above 0.34 for all the Swerling target models and for all values of n (and P_N).

4. The combination of fluctuation and integration losses is less than and decrease relative to the constant amplitude case as P_D decreases below 0.34 for all the Swerling target models and for all values of n (and P_N).
5. The combination of fluctuation and integration losses relative to the constant amplitude case for $P_D \neq 0.34$ is the highest for the Swerling I case; next highest for the Swerling III case; and much less for the Swerling II and IV cases with Swerling IV being the least and all are relatively insensitive to the value of n.
6. The combination of fluctuation and integration losses of the Swerling II and IV cases relative to the constant amplitude case approaches zero for all values of P_D as n approaches zero for all values of P_D as n approaches infinity. The three cases are practically identical for $n \geqslant 100$.
7. The integration loss (Table 4.4) increases as P_D decreases and n (and P_N — not demonstrated) increase for all five target models.

The integration loss varies most as a function of P_D for the Swerling II and IV cases. At high P_D there is actually an integration gain for these target models.

Cumulative Probability of Detection — Search radar designs are often based on specified cumulative detection probability. It is the probability that a target is detected at least once in M scans. This is to be distinguished from the probability of detection during a single scan.

$$P_C = 1 - \prod_{i=1}^{M} (1 - P_{Di}) \tag{4.44}$$

where

P_C = cumulative probability of detection for M scans

P_{Di} = probability of detection on the ith scan

M = total number of scans that the target can be detected.

In a typical situation, the search radar provides coverage of a volumetric sector in frame time, T. It is desired to achieve a specified P_C and P_N by the time a target reaches a given range R_o. The target is approaching the radar at its maximum constant radial velocity, v. The target first appears on the radar horizon at a maximum range, R_m.

Therefore

$$M = \frac{R_m - R_o}{\Delta R} \qquad (4.45)$$

Where: $\Delta R = vT$

and:

$$(S/N)_{i_{dB}} = 10 \lg (S/N)_{R_m} + 40 \lg \frac{R_m}{R_i} + \Delta L_{AT_{i_{dB}}} \qquad (4.46)$$

where:

$(S/N)_{i_{dB}}$	= S/N at range R_i in decibels
$(S/N)_{R_m}$	= S/N at range R_m — design value
R_m	= maximum unambiguous range
R_i	= $R_m - (i-1)\,\Delta R$
	= range on i^{th} scan
$\Delta L_{AT_{i_{dB}}}$	= the decrease in the two way propagation loss in decibels between ranges R_m and R_i

The following is a typical design procedure:

1. Develop search pattern and determine frame time, T.
2. Select $S/N_{RM} = S/N_1$ and determine P_{D1} (from graphs for the specified P_N and target fluctuation model). (or select P_{D1} and determine S/N_1)
3. Determine range increment between scans, ΔR, and total number of scans, M. (Given maximum target velocity, v. Maximum range, R_m, and detection range, R_o.)
4. Determine S/N_2, S/N_3, . . . S/N_M using Eq. (4.46) and ΔL_{AT_1}, ΔL_{AT_2}, . . . ΔL_{AT_m} from charts.
5. Determine P_{D_2}, P_{D_3}, . . . P_{D_M} from graphs for specified P_N and target fluctuation model.
6. Obtain P_C using Eq. (4.44).

If P_C is too small or too large repeat steps 2 through 6 assigning a larger or smaller value respectively of S/N_1 (or P_D). Several iterations may be required before a suitable set of design parameters is obtained.

An illustrative example has the following typical parameters:

R_m = 260 nm

v = MACH 2 = .367 nm/s

T = 8.17 s

P_N = .866 x 10^{-6}

$(S/N)_1$ = 6.0 dB

The progressive P_C is obtained using above parameters for all the target fluctuation models until $P_C > 0.99$ in all cases. The radar design is based on single hit detection on each scan. The effect of change in propagation loss has not been included in this example. The results are summarized in Table 4.5 and graphed in Fig. 4.8.

This example demonstrates the important point that a search radar design requiring a high P_C with a relatively large value of M is not influenced appreciably by the target fluctuation model. This result is to be expected since a large M leads to a distribution of the individual P_D values below and above 0.34 so that the fluctuation gains and losses respectively of the various target models tend to balance each other in the P_C process. In this example there is only a 0.5 dB spread for the S/N design requirements (11.0 to 11.5 dB) of all the target models to achieve a P_C = 0.99 ($M \approx 23$). This spread increases to 1.2 dB for the S/N design requirements (9.4 to 10.6 dB) to achieve a P_C = 0.90 ($M \approx 20$).

From another point of view the "saving" in power-aperture of designing to a high P_C instead of to a single scan P_D of the same value is appreciable and is significantly affected by the target model selected. For the above example:

Probability of detection = 0.90	Power-Aperture Saving dB
Constant Amplitude (13.2 — 10.6)	2.6
Swerling III, IV (17.3 — 9.6)	7.7
Swerling I, II (21.1 — 9.4)	11.7

Probability of detection = 0.99	Power-Aperture Saving dB
Constant Amplitude (14.5 — 11.5)	3.0
Swerling III, IV (23.0 — 11.0)	12.0
Swerling I, II (31.4 — 11.0)	20.4

Table 4-5
Cumulative Probability of Detection for Swerling and Constant Target Models

M	R nm	S/N dB	Constant P_D	Constant P_C	SW I, II P_D	SW I, II P_C	SW III, IV P_D	SW III, IV P_C
1	260	6.0	.010	.0100	.058	.0580	.040	.0400
2	257	6.2	.012	.0219	.062	.1180	.048	.0861
3	254	6.4	.014	.0356	.072	.1815	.056	.1380
4	251	6.6	.016	.0510	.080	.2475	.062	.1915
5	248	6.8	.018	.0681	.085	.3115	.066	.2450
6	245	7.0	.023	.0895	.092	.3750	.070	.2975
7	242	7.3	.030	.1168	.110	.4440	.088	.3595
8	239	7.5	.035	.1477	.118	.5090	.100	.4230
9	236	7.7	.040	.1818	.128	.5715	.110	.4860
10	233	7.9	.047	.2227	.135	.6295	.120	.5480
11	230	8.1	.058	.2678	.150	.6855	.130	.6070
12	227	8.4	.070	.3191	.160	.7359	.142	.6630
13	224	8.6	.080	.3736	.173	.7815	.165	.7185
14	221	8.8	.100	.4362	.190	.8231	.180	.7693
15	218	9.1	.115	.5110	.210	.8602	.200	.8157
16	215	9.3	.140	.5709	.225	.8917	.220	.8561
17	212	9.6	.175	.6460	.250	.9175	.255	.8929
18	209	9.8	.200	.7168	.265	.9403	.280	.9229
19	206	10.0	.250	.7876	.275	.9567	.297	.9458
20	203	10.3	.300	.8513	.300	.9697	.335	.9639
21	200	10.6	.360	.9048	.320	.9794	.358	.9768
22	197	10.8	.400	.9429	.330	.9862	.380	.9856
23	194	11.1	.460	.9692	.360	.9912	.400	.9914
24	191	11.4	.540	.9858	.380	.9945	.440	.9952
25	188	11.6	.580	.9940				

Target Velocity = MACH 2
Frame Time = 8.17 s

$P_N = .866 \times 10^{-6}$

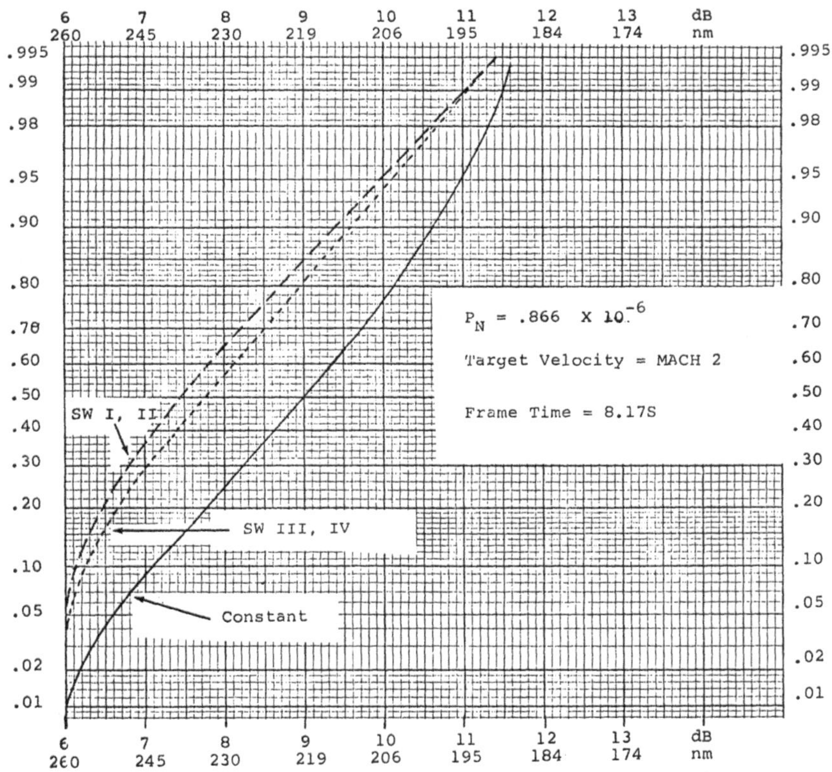

Figure 4-8
Cumulative P_D vs Range & S/N

References

[1] Marcum, J.I., "A Statistical Theory of Target Detection by Pulsed Radar, Mathematical Appendix," *Rand Report RM-753*, July 1, 1948; also in a special issue of *IRE Trans.*, "Prof. Group on Information Theory," Vol. IT-6, pp 145-267, April 1960.

[2] Swerling, P., "Probability of Detection for Fluctuating Targets," *Rand Report RM-1217*, March 17, 1954; also in a special issue of *IRE Trans.*, "Prof. Group on Information Theory," Vol. IT-6, pp 269-308, April 1960.

[3] DiFranco & Rubin, *Radar Detection*, Prentice-Hall, 1968.

[4] Meyer and Mayer, *Radar Detection Theory*, Academic Press, 1973.

List of Symbols

$C_{IF}(-p)$ $= \int_{o}^{\infty} p_1(R)_F \exp(-pR_{dR})$

$C_{oc}(-p)$ $= \int_{o}^{\infty} p(R)_c \exp(-pR_{dR})$ Fourier Transform

$C_{On}(-p)$ $= [C_{OC}(-p)]^n$ [Characteristic Function]

C_α = cost of Type I error (positive quantity)

C_β = cost of Type 2 error (positive quantity)

$I_o \dfrac{SR}{N}$ = modified Bessel function — results from averaging (integrating) out the random phase variation.

K = real positive number dependent on specification of α

L_A = average loss

L_L = integration loss relative to coherent integration

$L(y)$ = likelihood ratio

M = total number of scans in which the target can be detected

N = average noise power (mean-square noise voltage)

P_o = total probability of error

P_C = cumulative probability of detection for M scans

P_D = probability of detection

P_N = probability of false alarm

P_{D_i} = probability of detection on the ith scan

P_{OD} = single hit probability of detection for constant amplitude target model — Case O

P_{ON} = single hit false alarm rate for constant amplitude target model — Case O

$P_o(y)$ = probability of y with no signal present conditional probability

$P_s(y)$ = probability of y given s conditional probability

P_{OD_n} = probability of detection of constant amplitude target model for n hits — Case 1

P_{ON_n} = false alarm rate of constant amplitude target model for n hits — Case 1

P_{1D} = single hit probability of detection for SW I — Rayleigh target model — Case 1

P_{1N} = single hit false alarm rate for SW I — Rayleigh target model — Case 1

P_{1D_n} = probability of detection for SW I — Rayleigh target model for n hits — Case 1

P_{1N_n} = false alarm rate for SW I — Rayleigh target model for n hits — Case 1

P_{2D_n} = probability of detection for SW II — Rayleigh target model for n hits — Case 2

P_{2N_n} = false alarm rate for SW II — Rayleigh target model for n hits — Case 2

R = signal + noise - voltage

R_i = R_m -(i-1)ΔR = range on the ith scan

R_o = detection has to be achieved by this range

R_m = maximum unambiguous range

S = signal voltage

$(S/N)_i$ = S/N required to produce the same P_D and P_N for n integrated hits

$(S/N)_{i_{dB}}$ = S/N at range R_i in decibels

$(S/N)_{R_m}$ = S/N at range R_m — design value

$(S/N)_{SH}$	= single hit S/N required to produce a given P_D and P_N
T	= threshold voltage level
n	= number of hits integrated non-coherently
p	= a priori probability that signal is present
$p(R)_c$	= Rician pdf for envelope distribution of narrowband modulated sinusoidal signal of constant amplitude in the presence of noise for single hit
$p(R)_F$	= pdf for fluctuation target model in presence of noise for single hit
$p_1(R)_F$	= pdf for Rayleigh target cross section model in presence of noise for single hit — Case 1
$p_1(\sigma)$	= pdf for Rayleigh target cross section model
	$= \frac{1}{\sigma_o} \exp\left(\frac{-\sigma}{\sigma_o}\right)$
$p_2(\sigma)$	$= p_1(\sigma)$
$p_3(\sigma)$	$= \frac{4\sigma}{\sigma_o^2} \exp\left(\frac{-2\sigma}{\sigma_o}\right)$
	= pdf for One Dominant + Rayleigh target cross section model
$p_4(\sigma)$	$= p_3(\sigma)$
$p_5(\sigma)$	= pdf for Log-Normal target cross section model
$p_6(\sigma)$	= pdf for Weibull target cross section model
$p_{On}(R)_c$	$= \int_{-\infty}^{\infty} C_{ON}(-p) \exp(pR_{dp})$ Inverse Fourier Transform
	= pdf of constant amplitude target model in presence of noise for n hits — Case 1
$p_{1n}(R)_F$	= pdf for SW I — Rayleigh target model in presence of noise for n hits — Case 2
$p_{2n}(R)_F$	= pdf for SW II — Rayleigh target model in presence of noise for n hits — Case 2
q	= a priori probability that there is no signal
y	= S+N signal-plus-noise N noise alone continuous random variable

y_b $= \frac{T}{\sqrt{2N}}$

$\Delta L_{AT_{i_{dB}}}$ = the decrease in the two way propagation loss decibels between ranges R_m and R_i

ΔR = vT = radial target range increment per search frame time, T

α = probability of false alarm, P_N

β = probability that signal is absent, $(1 - P_D)$

σ_o = average cross section.

Matched Filter Theory and Waveforms

Chapter Five

5.1 Matched Filter Theory* — *Definition* — A filter network whose frequency response maximizes the output signal-to-noise ratio, S/N, is called a matched filter.

Where:

$$\frac{S}{N} = \frac{\text{peak instantaneous output signal power}}{\text{r m s output noise power}} \tag{5.1}$$

The matched filter is one step in the optimum linear processing for the detection of radar signals in noise.

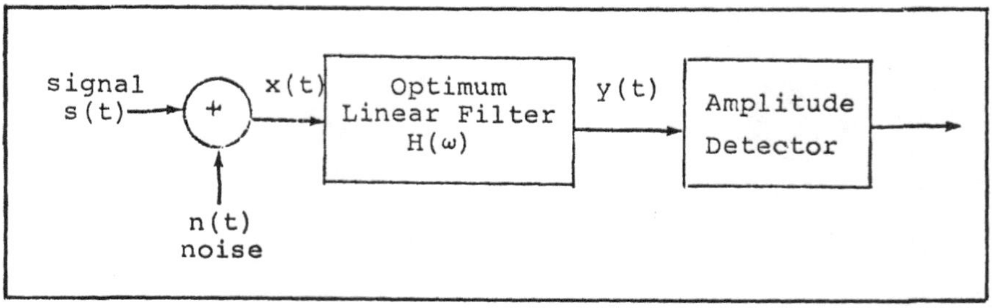

*North[1] is credited with deriving the properties of the optimum receiver for a signal in white gaussian noise.

One of the important tasks of a radar system designer is to select the optimum waveform for use in a radar system design to satisfy performance requirements in a particular function. The matched filter design is then determined to process the selected waveform. Subsequently a practical and realizable receiver and signal processor are designed (predetection r.f. and i.f. network stages) with a transfer function that is a reasonable approximation to the matched filter. The filter mismatch loss (S/N obtainable at the output of the matched filter divided by the S/N obtainable at the output of the realizable receiver/signal processor) for each selected waveform is determined and added to the system loss budget for use in sizing the radar system in terms of power-aperture requirements.

Optimum waveforms are not uniquely definable in the abstract. Waveform selection involves many factors. The most prominent considerations are accuracy, resolution, and ambiguities. Related considerations include: target density and dynamic range of radar target cross sections; weather and clutter; realizability of transmit waveforming networks and receive waveform processing networks; and such characteristics of transmitter candidates as peak power, average power, energy, short and long duty cycle, stability, amplitude, frequency, and phase.

The basic concept of matched filtering evolved from theoretical investigations of factors leading to optimum performance of radar systems. An extensive excellent coverage of this subject is contained in Refs. 2, 3, and 4. A derivation of matched filter requirements is presented to provide deeper insight into matched filter concepts and how they relate to radar system design.

If s(t) is the time domain representation of any physical waveform and $S(\omega)$ is the corresponding representation of this waveform in the frequency domain, then a filter which is matched to **s(t)** can be expressed either as a frequency or a time response function the two being interrelated by the Fourier transform:

$$S(\omega) = \int_{-\infty}^{\infty} s(t)e^{-j\omega t}\, dt \text{ (Fourier Transform)} \tag{5.2}$$

and

$$s(t) = \frac{1}{2\pi}\int_{-\infty}^{\infty} S(\omega)e^{j\omega t}\, d\omega \text{ (Inverse Fourier Transform} \tag{5.3}$$

A filter matched to s(t) is by definition one with impulse response

$$h(\tau) = a\, s(t_d - \tau) \tag{5.4}$$

a = arbitrary amplitude constant
t_d = arbitrary time delay constant

The transfer function of the matched filter is

$$H(\omega) = \int_{-\infty}^{\infty} h(\tau) e^{-j\omega\tau}\, d\tau \tag{5.5}$$

$$= a \int_{-\infty}^{\infty} s(t_d - \tau) e^{-j\omega\tau}\, d\tau \tag{5.6}$$

and letting $t' = t_d - \tau$

$$H(\omega) = a\, e^{-j\omega t}\, d \int_{-\infty}^{\infty} s(t') e^{j\omega t'}\, dt' \tag{5.7}$$

Comparison of equations (5.2) and (5.7) gives

$$H(\omega) = a\, S(-\omega) e^{-j\omega t_d} \tag{5.8}$$

or

$$H(\omega) = a\, S^*(\omega)\, e^{-j\omega t_d} \tag{5.9}$$

where:

$S^*(\omega)$ = complex conjugate of $S(\omega)$

∴ The transfer function of a matched filter is the complex conjugate of signal spectrum.

Derivation of the Maximum S/N Attainable at the Output of a Matche Filter —

Let x(t) of Fig. 5.1 be the received waveform.

where:

$$x(t) = \begin{cases} n(t) & \text{white noise alone} \\ \text{or} & \\ n(t) + s(t) & \text{white noise plus a signal return of known waveform} \end{cases}$$

Since the filter is linear, the output will contain a noise component, N, due to n(t) only and, if a signal is present, a signal component, S, due to s(t). To maximize S/N the filter must make the instantaneous power, S, as large as possible compared to the average noise power, N.

The average noise power in n(t) at any instant, assuming n(t) is stationary, is the integrated power under the noise power density spectrum at the filter output. If the input noise power density (power per unit bandwidth) is $N_o/2$ watts/hertz, the output noise power density is $(N_o/2)\,|H(\omega)|^2$ and the output average noise power is

$$N = \frac{N_o}{4\pi}\int_{-\infty}^{\infty} |H(\omega)|^2\, d\omega \qquad (5.10)$$

If $S(\omega)$ is the input signal spectrum, then $S(\omega)\,H(\omega)$ is the output signal spectrum and $s_o(t_d)$ is the inverse Fourier transform representing the output signal voltage at $t = t_d$ when it is at its maximum.

$$\therefore\ s_o(t_d) = \frac{1}{2\pi}\int_{-\infty}^{\infty} S(\omega)\, H(\omega) e^{j\omega t_d}\, d\omega \qquad (5.11)$$

and the output signal power (into one ohm) is:

$$S = \left[\frac{1}{2\pi}\int_{-\infty}^{\infty} S(\omega)\, H(\omega)\, e^{j\omega t_d}\, d\omega\right]^2 \qquad (5.12)$$

Therefore, the S/N to be maximized is:

$$\frac{S}{N} = \frac{2\left[\dfrac{1}{2\pi}\displaystyle\int_{-\infty}^{\infty} S(\omega)\, H(\omega)\, e^{j\omega t_d}\, d\omega\right]^2}{\dfrac{N_o}{2\pi}\displaystyle\int_{-\infty}^{\infty} |H(\omega)|^2\, d\omega} \qquad (5.13)$$

Since the integral in the numerator is real [output signal $s_o(t_d)$] the Schwarz inequality

$$\left|\int f(x)\, g(x)\, dx\right|^2 \leqslant \int |f(x)|^2\, dx \int |g(x)|^2\, dx \qquad (5.14)$$

can be used.

$$\therefore \text{Let } f(x) = H(\omega)$$
$$g(x) = S(\omega)\, e^{j\omega t_d}$$

$$\therefore \left(\frac{S}{N}\right)_{max} \leqslant \left[\frac{\frac{1}{\pi} \int_{-\infty}^{\infty} |H(\omega)|^2 \, d\omega \cdot \frac{1}{2\pi} \int_{-\infty}^{\infty} |S(\omega)|^2 \, d\omega}{\frac{N_o}{2\pi} \int_{-\infty}^{\infty} |H(\omega)|^2 \, d\omega} \right]$$

or

$$\left(\frac{S}{N}\right)_{max} \leqslant \left[\frac{1}{\pi N_o} \int_{-\infty}^{\infty} |S(\omega)|^2 \, d\omega \right] \tag{5.15}$$

From Parseval's Theorem —
Total signal energy is the same — whether obtained by integrating instantaneous power over all time or by integrating energy spectrum over all frequencies.

$$\frac{1}{2\pi} \int_{-\infty}^{\infty} |S(\omega)|^2 \, d\omega = \int_{-\infty}^{\infty} |s(t)|^2 \, dt = \text{signal energy} = E \tag{5.16}$$

then

$$\frac{S}{N} \leqslant \frac{2E}{N_o} \tag{5.17}$$

It is evident that Schwarz's inequality, and hence equations (5.15) and (5.17) apply when $f(x) = K\, g^*(x)$ or when

$$H(\omega) = a\, S^*(\omega) e^{-j\omega t_d} \tag{5.18}$$

Therefore, when a filter is matched to the received signal, s(t), a maximum value of S/N is obtained.

It can further be shown that the inequality holds only when $f(x) = K\, g^*(x)$ so the matched filter represents the only type of linear filter which maximizes S/N.

The radar system designer infers from equation (5.17) that, with matched filter receiver processing, the ability to detect a radar signal return is only a function of its energy content and is completely independent of the signal peak power, waveshape, time duration, and bandwidth.

5.2 **Filter Mismatch Loss** — When the realizable receiver filter, consisting of the i.f. amplifier or i.f. waveform processor, is not matched to the signal spectrum, the output i.f. signal-to-noise ratio will be less than from an ideal matched filter. The loss is due to exclusion of signal from the filter pass band and/or inclusion of excess noise in the filter pass band. The filter mismatch loss, L_M, is the ratio of the signal-to-noise ratio from a matched filter, $(S/N)_{MF}$, to the signal-to-noise ratio from a nonmatched filter, $(S/N)_{NMF}$.

$$L_M = \frac{(S/N)_{MF}}{(S/N)_{NMF}} \tag{5.18}$$

or using equation (5.17) for (S/N)

$$L_M = \frac{2\,E/N_o}{(S/N)_{NMF}} \tag{5.19}$$

and equation (5.13) with t replacing t_d for the general case of S/N_{NMF}

$$L_M = \frac{\dfrac{E}{2\pi}\displaystyle\int_{-\infty}^{\infty} |H(\omega)|^2\, d\omega}{\left|\dfrac{1}{2\pi}\displaystyle\int_{-\infty}^{\infty} S(\omega)\, H(w)\, e^{j\omega t}\, d\omega\right.^2 \quad \text{Maximized}} \tag{5.20}$$

where:

$H(\omega)$ = transfer function of the mismatched filter.

The value of the output signal power is obtained for the time at which it is a maximum.

Equation (5.16) obtains the general form

$$L_M = \frac{\int_{-\infty}^{\infty} |S(\omega)|^2 \, d\omega \cdot \int_{-\infty}^{\infty} |H(\omega)|^2 \, d\omega}{\left| \int_{-\infty}^{\infty} S(\omega) \, H(\omega) \, e^{j\omega t} d\omega \right|^2 \text{ Maximized}} \tag{5.21}$$

or in the time domain

$$L_M = \frac{\int_{-\infty}^{\infty} |s(t)|^2 \, dt \cdot \int_{-\infty}^{\infty} |h(t)|^2 \, dt}{\left| \int_{-\infty}^{\infty} s(\tau) \, h(t-\tau) \, d\tau \right|^2 \text{ Maximized}} \tag{5.22}$$

5.3 Types of Waveforms for ESR Applications — In practically all radar applications the type of waveforms employed have rectangular envelopes. Most transmitters perform best with rectangular envelope waveforms. They operate more efficiently and provide good output amplitude stability by operating at their peak power saturation level. Rectangular envelope waveforms may be grouped into the following three categories:

Monotonic — constant carrier frequency

Frequency Modulated Coded —
- linear fm, parabolic fm, cosine fm } analog coded
- discrete frequency sequence, binary phase } discrete coded

Pulse Trains —
- uniform or staggered spacing
  - monotonic
  - fm coded

In the following discussion it is assumed the reader is familiar with these waveforms. Their detailed characteristics are given in Refs. 4, 5 and 6. All three types are used in ESR applications with fm coded waveforms being the most common and pulse trains the least.

In a large number of ESRS s, the simpler single beam design (instead of multiple simultaneous beams) is employed for multimode operation on a time-shared basis. Radar time is usually at a premium in these designs and time is conserved by transmitting only a single pulse per beam position in the search and track operational modes. The fm coded waveforms are very suitable for these applications. The linear fm coded waveform being the most common. The relatively high energy content in each pulse (compared to multiple pulse integration) is obtained for practical peak power limits by using long pulse durations. By manipulation of the following simple formulas the ESRS designer has great flexibility in the use of energy, peak power, pulse duration and signal bandwidth.

$$E = P_p T \tag{5.23}$$

$$TB = T/\tau = T\Delta f$$

where:

- E = energy — joules
- P_p = peak power — watts
- T = pulse duration — s
- B = signal bandwidth — hertz
- Δf = linear frequency modulation in time T — hertz
- τ = compressed pulse width — seconds
- TB = time — bandwidth product
- T/τ = compression ratio

For the various reasons given in Chapter 10, it is desirable to employ energy management in the real time control of an ESRS by adjustment of the peak power and/or pulse duration. Peak power control on most transmitters is usually not available since transmitters are mostly peak power limited. A two level control is obtainable in some types of transmitters (those using a CFA output tube for example) by turning off the output tube and allowing the driver stage power to pass through it to the antenna. The discrete step power reduction obtained is equal to the gain of the output tube plus the passive losses through the output tube. Pulse duration adjustment of fm coded waveforms is employed more extensively because it provides a wider range of control for energy management. The pulse duration is varied as a linear function of energy. Both the pulse-length and the time-bandwidth product are adjusted in binary increments to maintain the same compressed pulse width. The ESRS designer often maintains the same range resolution (signal bandwidth) for the search and track functions because of the following advantages:

No time is lost nor implementation required to perform a range cell search in handing a target over from normal or MTI search to target acquisition.

There is less likelihood of multiple target ambiguity in handover from one coverage sector to another.

Receiver bandwidth is the same for all waveforms and operational modes thus simplifying implementation and threshold level setting.

Binary incremental control of the pulse length and time-bandwidth product simplifies the design of the signal processor. (In an analog or digital signal processor the various waveforms are obtained by switching in-or-out binary segments of the matched filter network.)

A lower limit of time-bandwidth product is approximately 10. It is difficult to achieve acceptable sidelobe levels for small time-bandwidth products. If the energy requirement decreases substantially (as occurs in many MTI applications where short ranges are involved) the lower limit waveform can consist of a monotonic or single frequency pulse with a pulse width equal to the reciprocal of the signal bandwidth of the fm coded waveforms. The usable maximum pulse duration is limited by transmitter capability and/or specified minimum range coverage. The minimum range restriction can be circumvented by a second coverage of the search sector with short pulses having short dwell periods. The penalty paid is increased time usage. The upper limit of time-bandwidth product achievable with existing technology is extremely high. Time-bandwidth products in the tens of thousands have been obtained and even higher in some unique designs. Typical values are in the 50 to 200 range.

A monotonic or single frequency waveform is the simplest to generate and process. It is used whenever the energy, peak power, pulse duration, and resolution (or signal bandwidth) performance requirements can be mutually satisfied; i.e., when the pulse duration meets the resolution requirements and there is adequate peak power available to also satisfy the energy requirements of the radar system design.

Pulse trains, either monotonic or fm codes, are employed in ESRS s for two purposes: to determine doppler velocities and fluctuation characteristics of objects. Both of these applications require a finite time in one beam position while a train of pulses is transmitted and received. Typical values range from 1 to 20 milliseconds for the pulse train length and from 10 to 500 microseconds for the interpulse period. Long fm coded waveforms are occasionally used to obtain doppler data on targets if they are consistent with the energy, peak power and pulse duration capabilities of the transmitter and the radar system parameters.

In many applications the presence of the large range sidelobes or time residues of all these waveforms limit the capability of the radar to examine a dense multiple target environment over an extended dynamic range of target radar cross section. The effects of the time residues are minimized by the use of time domain envelope shaping of the transmitted signal or by frequency domain weighting of the receive response network or a combination of both. In most cases the weighting is performed in the receiver to retain peak power limited operation in the transmitter. The radar system designer considers an optimum weighting function one that minimizes the mismatch and resolution losses for a maximum reduction in sidelobe level. In some applications the rate at which range sidelobes fall off may be an important consideration; especially if in accepting somewhat larger sidelobes adjacent to the main peak of the pulse there is less mismatch and resolution loss. In general, the radar system designer is most likely to select the weighted mismatch filter design that reduces the peak sidelobe (usually the first) to the level required to meet the target cross section dynamic range specification, for the minimum mismatch loss. The pulse widening introduced by the selected weighting function is compensated for by a corresponding increase in signal bandwidth over what is required to meet the resolution specification with a matched filter. For a range of peak sidelobes from 23 to 40 dB the mismatch losses range from 0.6 to 2.4 dB and the pulse widening from 1.1 to 1.9. These parameters are obtained, for various filter networks and weighting functions for the waveforms discussed, in Refs. 3 through 8. Tabulations are given on Ref. 3, Page 414; Ref. 4, Page 205; and Ref. 8, Pages 20 through 30.

References

[1] D.O. North, "Signal/Noise Discrimination in Pulse Carrier Systems", RCA Laboratories, Princeton, NJ, Tech. Rpt. No. PTR-6C, June 25, 1943. Republished *Proc. IEEE*, July 1963, pp 1016-1027.

[2] G.L. Turin, "An Introduction to Matched Filters", *IRE Trans. Information Theory*, Vol. IT-6, June 1960, pp 311-329.

[3] M.I. Skolnik, *Introduction to Radar Systems*, McGraw-Hill, 1962.

[4] C.E. Cook & M. Bernfeld, *Radar Signals*, Academic Press, 1967.

[5] Difranco & Rubin, *Radar Detection*, Prentice Hall, 1968.

[6] A.W. Rihaczek, "Radar Resolution Properties of Pulse Trains", *Proceedings of the IEEE*, February 1964, pp 153-164.

[7] Barton & Ward, *Handbook of Radar Measurements*, Prentice Hall, 1969.

[8] M.I. Skolnik, *Radar Handbook*, McGraw-Hill, 1970.

List of Symbols

B	= signal bandwidth — hertz
E	= energy — joules
$H(\omega)$	= filter frequency spectrum — transfer function of filter
L_M	= mismatch filter loss
N_o	= noise power density
P_p	= peak power — watts
$\frac{S}{N}$	$= \frac{\text{peak instantaneous output signal power}}{\text{rms output noise power}}$
$(S/N)_{MF}$	= signal-to-noise ratio from a matched filter
$(S/N)_{NMF}$	= signal-to-noise ratio from a non-matched filter
$S(\omega)$	$= \int_{-\infty}^{\infty} s(t)e^{-j\omega t}\,dt$ (Fourier Transform)
$S^*(\omega)$	= complex conjugate of $S(\omega)$
T	= pulse duration — s
TB	= time-bandwidth product
T/τ	= compression ratio
a	= arbitrary constant
$h(\tau)$	= impulse response

n(t)	= white noise
s(t)	$= \frac{1}{2\pi} \int_{-\infty}^{\infty} S(\omega) e^{j\omega t} \, d\omega$ (Inverse Fourier Transform)
	= physical waveform in time domain
t_d	= arbitrary time delay constant
x(t)	= received waveform
y(t)	= received waveform after optimum linear processing
Δf	= linear fm in time T — hertz
τ	= compressed pulse width — s

MTI Consider-ations

Chapter Six

6.1 Definition — MTI stands for moving target indication. The doppler shift in frequency caused by a moving target permits the pulse radar to discern moving targets in the presence of fixed targets. The target echo of a moving target competes with target echoes of fixed targets (which do not have doppler shifts) in the same radar resolution cell. The fixed targets enter this radar resolution cell if they are at the same range as the moving target but not necessarily at the same angular position; the fixed target echoes may be received from the sidelobes as well as from the main lobe of the antenna. The fixed or slow moving targets to be eliminated by MTI can be ground clutter, sea clutter, rain, snow, or chaff.

A distinction is made between MTI and pulse doppler processing. The pulse doppler radar can provide both fixed target rejection and doppler (velocity) measurement of moving targets. MTI is only used for fixed target rejection. MTI systems are usually operated at relatively low pulse repetition rates to provide unambiguous range with ambiguous doppler while pulse doppler systems are usually operated at high pulse repetition rates to provide unambiguous doppler with ambiguous range. Consistent with this distinction, MTI radars employ delay line cancellers to eliminate the fixed targets by a sweep-to-sweep subtraction while pulse doppler radars usually employ range gates with each range gate containing a bank of filters. The pulse doppler systems are more complex to implement and require better stability than the MTI systems. The material covered in this chapter concerns MTI processing employing sweep-to-sweep cancellation. This is the technique employed in most conventional and electronic scanning radar applications for fixed or slow moving target rejection.

Descriptions of the various types of MTI's and their implementation have been well covered in the literature (Refs. 1, 2, 3, and 4). The intent here is to indicate the basic MTI design formulas and their usage by the ESRS designer. In the process, MTI characteristics unique to ESRS s will be indicated. The MTI design procedure as applied to ESRS s will be demonstrated by an illustrative example.

6.2 MTI Design Formulas — An important quantity which provides a measure of MTI performance is the MTI *improvement factor* (Ref. 5). It is defined as the signal-to-clutter ratio at the output of the MTI system compared with that of the input where the signal is averaged uniformly over all radial velocities.

The required improvement factor of an MTI system is dependent on the amount of clutter power relative to signal power intercepted by the radar resolution cell and the single-pulse detectability factor, $(S/N)_o$, (the value of S/N required to achieve a given P_D with fixed P_N) of the radar design as shown in Fig. 6.1. The required improvement factor is simply the amount by which the clutter has to be reduced so that the radar system detection capability is approximately the same as in the absence of clutter. An acceptable design goal is to provide approximately a 6 dB margin of the clutter residue below the noise level.

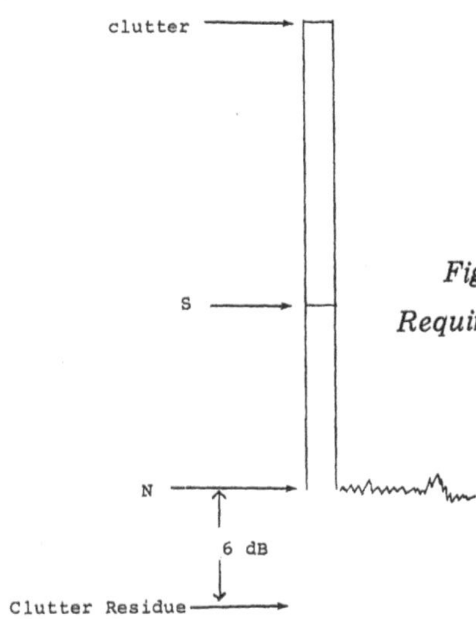

Figure 6-1

Required Improvement Factor

$$\therefore\ I_{R_{dB}} = (S/N)_{o_{dB}} + \frac{\sigma_c}{\sigma_{t_{dB}}} + 6\ \text{dB} \tag{6.1}$$

where:

- I_R = required improvement factor
- σ_c = input clutter cross section or input clutter power
- σ_t = radar target cross section or input target power

By definition:

$$I = \frac{S_o/C_o}{S_i/C_i} = \frac{S_o/S_i}{C_o/C_i} = \frac{C_i/C_o}{S_i/S_o} \tag{6.2}$$

where:

- I = improvement factor
- S_o/C_o = signal-to-clutter ratio at the output of the MTI system
- S_i/C_i = signal-to-clutter ratio at the input of the MTI system
- C_i/C_o = clutter attenuation factor due to MTI processing
- S_i/S_o = signal attenuation factor due to MTI processing

$$I = \frac{\text{clutter attenuation factor}}{\text{signal attenuation factor}} = \frac{\dfrac{\int_0^\infty C(\omega)\,d\omega}{\int_0^\infty C(\omega)\,|H(\omega)|^2\,d\omega}}{\left.\dfrac{\int_0^\infty S(\omega)\,d\omega}{\int_0^\infty S(\omega)\,|H(\omega)|^2\,d\omega}\right|_{\text{averaged over all radial velocities}}} \tag{6.3}$$

- $C(\omega)$ = clutter power spectrum
- $|H(\omega)|^2$ = MTI delay line canceller power transfer function
- $S(\omega)$ = signal power spectrum

The gaussian shape provides a reasonable description of the clutter spectrum (Refs. 3 and 6).

$$C(\omega) = \exp\left[-\frac{(f - f_m)^2}{2\sigma_p^2}\right] = \exp\left[-\alpha\frac{(f - f_m)^2}{f_o^2}\right] \quad (6.4)$$

where:

σ_p = standard deviation of clutter spectrum (frequency spread of clutter)
f_m = mean frequency of clutter spectrum
f_o = transmission frequency
α = Barlow's dimensionless stability parameter

$$= \frac{f_o^2}{2\sigma_p^2}$$

The power transfer function for an n-delay-line canceller is:

$$|H_n(\omega)|^2 = \sin^{2n}\frac{\pi f}{f_r} \quad (6.5)$$

n = 1, 2, 3 . . .
f_r = repetition frequency

The signal power spectrum for uniform radial velocity distribution is:

$$S(\omega) = 1 \qquad 0 \leqslant f \leqslant f_r \quad (6.6)$$

The improvement factors for different types of MTI cancellers and design values for the clutter parameters have been conveniently tabulated in Ref. 7 for the use of radar system designers. Extensive MTI design data is available in Refs. 1-7. Commonly used good approximations for single and double coherent cancellers improvement factors are given below in several forms:

$$I_{2p}^{-1} = \frac{9.9\, f_o^2\, t_o^2}{\alpha} \left[1 - \frac{4.95\, f_o^2\, t_o^2}{\alpha} \right] \tag{6.7}$$

or

$$\frac{79\sigma_v^2\, t_o^2}{\lambda_o^2} \left[1 - \frac{39.5\sigma_v^2\, t_o^2}{\lambda_o^2} \right]$$

or

$$\frac{20\sigma_v^2}{V_b^2} \left[1 - \frac{10\sigma_v^2}{V_b^2} \right]$$

$$I_{3p}^{-1} = \frac{195 f_o^4\, t_o^4}{\alpha^2} + \frac{13 f_o^2\, \delta^2}{\alpha} \tag{6.8}$$

or

$$\frac{1.25 \times 10^4\, \sigma_v^4\, t_o^4}{\lambda_o^4} + \frac{105\sigma_v^2\, \delta^2}{\lambda_o^2}$$

or

$$\frac{780\sigma_v^4}{V_b^4} + \frac{26\sigma_v^2\, \Delta^2}{V_b^2}$$

where:

I_{2p} = single canceller improvement factor
f_o = transmission frequency — hertz
t_o = interpulse period — seconds (delay line length)
α = Barlow's dimensionless stability parameter
λ_o = wavelength of transmission — meters
σ_v = velocity spread — m/s

$$= \frac{c}{\sqrt{8\alpha}} \quad (c = \text{velocity of light} - \text{m/s})$$

$$= \frac{\lambda_o}{2}\,\sigma_p \quad (\sigma_p = \text{frequency spread of clutter})$$

V_b = first blind speed — m/s

$$V_{b_n} = \frac{n\lambda_o}{2T} = \frac{n\lambda_o\, f_r}{2} \quad \begin{array}{l}(f_r = \text{pulse repetition rate} - \text{hertz})\\(n = 1, 2, \ldots\ldots.)\end{array}$$

I_{3_p} = double canceller improvement factor

δ = stagger interval — seconds

= Δt_o (Δ = fraction of interpulse period) (usually Δ = .05 to .1)

Antenna Motion Limitation — The clutter spectrum broadens due to angular rotation of the antenna because the radar does not receive echoes from the identical patch of scatterers from pulse-to-pulse. For a gaussian two-way antenna pattern:

$$\sigma_A = \frac{\omega_A}{3.78\theta_A} \tag{6.9}$$

where:

σ_A = clutter spectrum broadening due to antenna motion — hertz

ω_A = antenna angular rotation rate — radians/sec

θ_A = horizontal beamwidth (3 dB — value one way-radians)

The corresponding improvement factor limitation imposed by the antenna motion is

$$I_{A_{2p}} = \frac{n_o^2}{1.39} \tag{6.10}$$

and

$$I_{A_{3p}} = \frac{n_o^4}{11.5} \tag{6.11}$$

where:

$I_{A_{2p}}$ = improvement factor limitation imposed by antenna motion for single canceller

$I_{A_{3p}}$ = improvement factor limitation imposed by antenna motion for double canceller

n_o = number of hits in 3 dB beamwidth

$$= \frac{f_r\,\theta_A}{\omega_A} = \frac{\theta_A}{t_o\,\omega_A}$$

A unique advantage of step scanned ESRS s is that there is no degradation in MTI performance due to antenna motion. The beam is held stationary at each discrete beam positioning during the transmission and reception of a MTI pulse train.

Instability Limitations — The clutter spectrum is further spread by system instabilities such as pulse-to-pulse variations in: transmitter frequency, stalo or coho frequency, pulse width, pulse amplitude, etc. (see Ref. 8). The radar system designer accounts for contributions from all these sources, including antenna motion by combining the standard deviations of the contributing factors or the improvement factor terms as follows:

$$\sigma_T^2 = \sum_{i=1}^{n} \sigma_i^2 \tag{6.12}$$

and

$$\frac{1}{I_T} = \sum_{i-1}^{n} \frac{1}{I_i} \tag{6.13}$$

where:

σ_T = total clutter spectrum spread (σ_p or σ_v) — hertz

σ_i = clutter spectrum spread due to the ith contributor — hertz (pulse jitter for example)

I_T = the total improvement factor available from the radar

I_i = improvement factor term due to the ith contributor

The radar system designer can obtain satisfactory MTI performance only when his design meets the requirement

$$I_R \leqslant I_T \tag{6.14}$$

or

$$\left(\frac{S}{N}\right)_{o_{dB}} + \frac{\sigma_c}{\sigma_{t_{dB}}} + 6\text{ dB} \leqslant \frac{1}{\underset{\text{Processor}}{\frac{1}{I_{MTI}}} + \underset{\text{Motion}}{\frac{1}{I_{Antenna}}} + \underset{\text{System Instabilities}}{\Sigma\frac{1}{I_i}}} \tag{6.15}$$

The remaining term to be considered is the input clutter to target power (or cross section) ratio, σ_c/σ_t. The target radar cross section, σ_t, is usually specified for a radar design. The cross section of the distributed clutter is determined by multiplying the clutter reflectivity by the area or volume of the resolution cell at the maximum range at which the moving target is to be detected in clutter.

Clutter Area —

$$\sigma_c = \sigma_o A_c = \sigma_o \frac{c\tau}{2} \frac{R_m \theta_A \sec\xi}{L_{BS}} \tag{6.16}$$

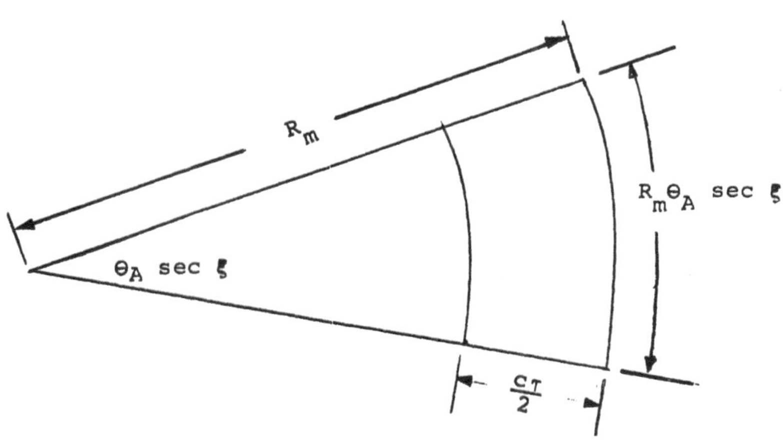

Clutter Volume —

$$\sigma_c = \eta V_c = \eta \frac{c\tau}{2} \cdot \frac{R_m^2 \, \theta_A \, \theta_E \sec\xi}{L^2_{BS}} \tag{6.17}$$

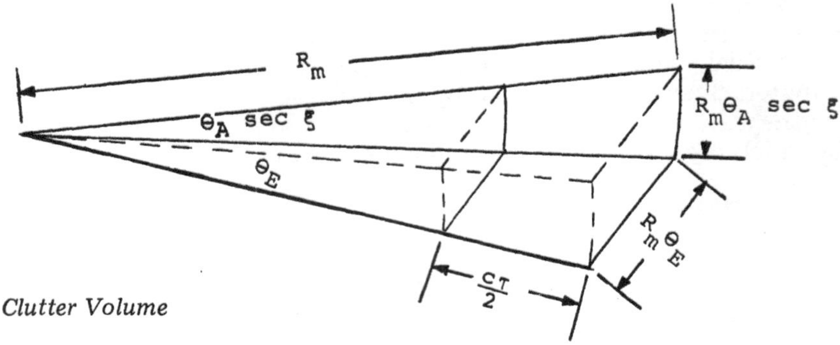

Clutter Volume

where:

σ_c	= clutter cross section — m^2
σ_o	= area clutter reflectivity (distributed land clutter or sea clutter distribution)
A_c	= area of distributed clutter resolution cell — m^2
c	= velocity of light — m/s
τ	= effective pulse width — seconds (reciprocal of signal bandwidth)
R_m	= maximum range at which moving target is to be detected in distributed clutter — m
θ_A	= azimuth 3 dB beamwidth — radians

θ_E = elevation 3 dB beamwidth — radians

ξ = scan angle from array normal at which electronic scan antenna beam intercepts clutter resolution cell-degrees (usually equal to maximum scan angle of electronic scan antenna)

L_{BS} = 1.33, beam shape loss

V_c = volume of distributed clutter resolution cell — m^3

η = volume clutter reflectivity — m^2/m^3

The sec ξ term in Eqs. (6.16) and (6.17) applies only for electronically scanned antennas. Since the antenna is stationary while the beam is being scanned, the effective aperture that contributes to the formation of the beam decreases as the cosine of the scan angle off broadside, ξ, as shown in Fig. 6.2. As the effective aperture decreases by cos ξ, the beam broadens by the reciprocal function sec ξ.

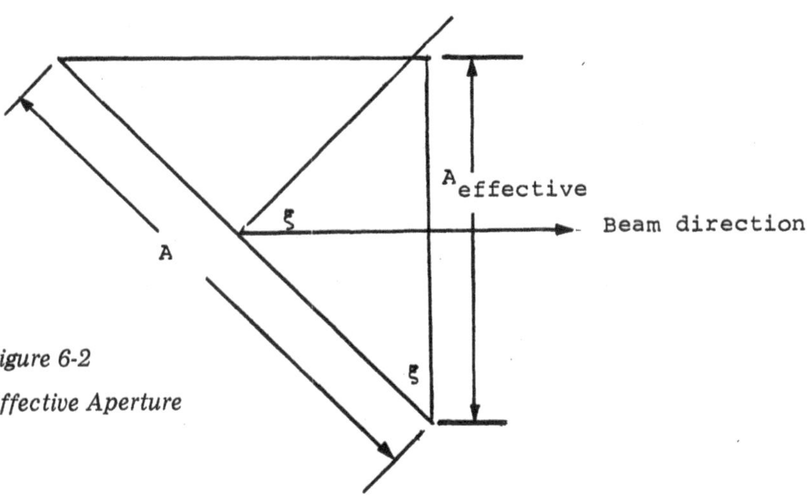

Figure 6-2

Effective Aperture

Expressions (6.16) and (6.17) can be expanded to include other considerations. For systems with range ambiguity or substantial second time around clutter contributions, the clutter competing with a moving target can be at a different range. In addition, targets at high elevation angles compete with clutter received from the antenna sidelobes. For this more general case

$$\frac{\sigma_c}{\sigma_T} = \sigma_o \frac{c\tau}{2} \frac{R_c \theta_A \sec\xi}{\sigma_T L_{BS}} \left(\frac{R_m}{R_c}\right)^4 G_{SL}^2 \qquad (6.16a)$$

and

$$\frac{\sigma_c}{\sigma_T} = \eta \frac{c\tau}{2} \cdot \frac{R_c^2 \theta_A \theta_E}{\sigma_T L_{BS}^2} \sec\xi \cdot \left(\frac{R_m}{R_c}\right)^4 G_{SL}^2 \qquad (6.17a)$$

where:

R_c = range of clutter — meters

R_m = range of target — meters

G_{SL} = antenna pattern sidelobe level illuminating clutter cell (less than unity)

Expressions (6.16a) and (6.17a) reduce to (6.16) and (6.17) for clutter located at the same range as the target ($R_c = R_m$) and received through the main lobe ($G_{SL} = 1$).

The clutter contributions from all the illuminated clutter cells can be obtained using the above expressions for each illuminated clutter cell with the corresponding values of G_{SL}, σ_o, and η. In many MTI designs the clutter area is evaluated over a broad ring of sidelobe response covering π radians in azimuth. The corresponding G_{SL} and σ_o terms used to determine the clutter area are obtained by averaging over the surface in front of the radar.

When the output of the MTI processor is being integrated and the clutter is partially correlated from pulse-to-pulse there will be an additional integration loss, term, L_{IC}, to introduce into the radar range equation to account for the reduced number of independent samples.

6.3 MTI Usage in ESRS Applications — The radar system designer can obtain considerable insight into the effect of radar parameters on MTI performance by consideration of expression (6.15). Assume system instabilities have been reduced to negligible values and there is no degradation in MTI performance due to antenna motion (step scan ESRS design). Substitution of equations (6.8), (6.16a), and (6.17a) into (6.15) and converting decibels to ratios obtains:

$$4\left(\frac{S}{N}\right)_o \sigma_o \frac{c\tau}{2} \frac{R_c \theta_A \sec\xi}{\sigma_T L_{BS}} \left(\frac{R_m}{R_c}\right)^4 G^2_{SL}$$

$$\leqslant \frac{1}{\dfrac{1.25 \times 10^4 \sigma_v^4 t_o^4}{\lambda_o^4} + \dfrac{105\, \sigma_v^2 \delta^2}{\lambda_o^2}} \tag{6.15a}$$

$$4\left(\frac{S}{N}\right)_o \eta \frac{c\tau}{2} \frac{R_c^2 \theta_A \theta_E \sec\xi}{\sigma_T L_{BS}^{\;2}} \left(\frac{R_m}{R_c}\right)^4 G^2_{SL}$$

$$\leqslant \frac{1}{\dfrac{1.25 \times 10^4 \sigma_v^4 t_o^4}{\lambda_o^4} + \dfrac{105\, \sigma_v^2 \delta^2}{\lambda_o^2}} \tag{6.15b}$$

From Table D-1, Ref. 7:

Reflectivity	Velocity Spread σ_v - m/s	Type of Clutter
$\sigma_o = \dfrac{.00032}{\lambda_o}$	.06 - .3	distributed land clutter / λ_o - m
$\sigma_{o_{dB}} = -64_{dB} + 6\, K_{B_{dB}} + \sin E_{dB} - \lambda_{o_{dBm}}$	.45 - 1.0	sea clutter K_B = 1,2,3,4,5 (sea state)(Beaufort scale) E = elevation angle λ_o - m
$\eta = 3 \times 10^{-8} \lambda_o$	.6 - 1.2	chaff (for fixed weight per unit volume) / λ_o - m
$\eta = \dfrac{6 \times 10^{-14}\, r^{1.6}}{\lambda_o^4}$	1.8 - 3.0	rain rate r - mm/hr λ_o - m

Recalling that the left side of Eq. (6.15, a & b) represents the required improvement factor and the right side the available improvement factor, it is obvious that the improvement factor of the MTI processor alone cannot indicate adequate MTI performance independent of the radar parameters. For example, if the effective pulse width is reduced from 1 μsec to 0.1 μsec the required improvement factor is reduced by 10 dB and consequently the requirements on the MTI processor are relaxed by 10 dB.

Improved MTI performance is obtained for a radar system design which has:

Low transmission frequency
(long transmission wavelength)

High pulse repetition rate (but no range ambiguities)
(short interpulse period)

Large signal bandwidth
(short effective pulse width)

Narrow beams

Low maximum and average antenna sidelobes

Low or no stagger on double or higher order cancellers

Step scanning

Small elevation tilt angle

For ESRS applications a finite MTI pulse train is transmitted and received at each discrete beam position and the processor is cleared (instead of operating continuously) for the next beam position. The number of pulses in the pulse train is dependent on the number of canceller stages and the amount of post MTI integration performed. In selecting the MTI processor it is useful to note that

$$I_{3p}\Big|_{\delta=0} \approx \tfrac{1}{2}\, I_{2p}^{\;2} \qquad (6.18)$$

For a double or higher order canceller the pulse repetition rate is usually staggered to fill in the blind speed notches. The staggering degrades the improvement factor obtainable from the canceller but can be kept within acceptable limits. However, a staggered pulse repetition rate cannot be used when there is a significant contribution from second time around clutter. With staggering, all second time around returns are displaced by the stagger increment, δ, on alternate pulses and consequently are not in time alignment for subtraction (cancellation). To counter second time around clutter two or more pulse repetition rates are employed to fill in the blind speed notches; and the unambiguous returns from the initial pulse, known as the "fill pulse", of each pulse train are blanked so that they do not contribute to the residue output from the MTI processor. An alternate or additional technique employed to eliminate blind speed notches is to transmit the second pulse train at a different frequency.

6.4 MTI Design — *Illustrative Example* — For a radar having the following parameters:

Transmission frequency, f_o	5 GHz
Antenna beamwidth, θ_A	1° pencil beam
Antenna scan angle, ξ	30°
Signal bandwidth, B	10 MHz
Detectability factor, $(S/N)_o$	11 dB
Target cross section, σ_t	1 m^2

select an MTI processor design to operate in a forest type of land clutter environment in the light rain (2 mm/hr) with the clutter extending in front of the radar to approximately the full target range of 130 Km.

Land Clutter

$$I_R = 4\left(\frac{S}{N}\right)_o \sigma_o \frac{c\tau}{2} \frac{R_c \theta_A \sec\xi}{\sigma_T L_{BS}}$$

$$\sigma_o = \frac{.00032}{\lambda_o} = \frac{.00032}{.06} = 5.3 \times 10^{-3}$$

$$\tau = \frac{1}{B} = \frac{1}{10^7} = 10^{-7}\,s$$

$$\left(\frac{S}{N}\right)_o = 11 \text{ dB} = 12.6$$

$$I_R = \frac{4 \text{x} 12.6 \text{x} 5.3 \text{x} 10^{-3} \text{x} 3 \text{x} 10^{8} \text{x} 10^{-7} \text{x} 1.3 \text{x} 10^{5} \text{x} .01745 \text{x} \sec 30^\circ}{2 \text{ x } 1 \text{ x } 1.33}$$

$$I_R = 7891.4 = 38.97 \text{ dB}$$

Single Canceller

$$I_A^{-1} = \frac{79\sigma_v^2 t_o^2}{\lambda_o^2}\left[1 - \tfrac{1}{2}\left(\frac{79\sigma_v^2 t_o^2}{\lambda_o^2}\right)\right]$$

$$\sigma_v = 0.3 \text{ m/s}$$

$$t_o = 867\ \mu\text{s} \quad (130 \text{ Km})$$
$$+133\ \mu\text{s} \quad (\text{retrace time})$$
$$1000\ \mu\text{s} = 10^{-3}\text{s}$$

$$\lambda_o = .06\text{m}$$

$$I_A^{-1} = \frac{79 \text{ x } .3^2 \text{ x } 10^{-6}}{.06^2}\left[1 - \tfrac{1}{2}(\quad)\right]$$

$$I_A^{-1} = .001975\left[1 - \frac{.001975}{2}\right]$$

negligible

$$I_A = 27.0 \text{ dB} \quad \text{inadequate}$$

Double Canceller

$$I_A^{-1} = \frac{1.25 \times 10^4 \sigma_v^4 t_o^4}{\lambda_o^4} + \frac{105\, \sigma_v^2 \delta^2}{\lambda_o^2}$$

$$\delta = .05\, t_o = .05 \times 10^{-3} = 5 \times 10^{-5}\,\text{s}$$

$$I_A^{-1} = \frac{1.25 \times 10^4 \times .3^4 \times 10^{-12}}{.06^4} + \frac{105 \times .3^2 \times (5 \times 10^{-5})^2}{.06^2}$$

$$I_A^{-1} = 7.81 \times 10^{-6} + 6.56 \times 10^{-6} = 1.437 \times 10^{-5}$$

$$I_A = 48.4 \text{ dB} \qquad \text{OK}$$

Rain — 2 mm/hr.

$$I_R = 4\left(\frac{S}{N}\right)_o \eta \frac{c\tau}{2} \quad \frac{R_c^2\, \theta_A\, \theta_E \sec\xi}{\sigma_T L_{BS}^2}$$

$$\eta = \frac{6 \times 10^{-14}\, r^{1.6}}{\lambda_o^4} = \frac{6 \times 10^{-14} \times 2^{1.6}}{.06^4} = 1.4 \times 10^{-8}$$

$$I_R = \frac{4 \text{x} 12.6 \text{x} 1.4 \text{x} 10^{-8} \text{x} 3 \text{x} 10^8 \text{x} 10^{-7} \text{x} 10^{10} \text{x} .01745^2 \text{x} 1.155}{2 \times 1 \times 1.33^2}$$

$$I_R = 21.0 = 13.25 \text{ dB}$$

Single Canceller

$$\sigma_v = 2.0 \text{ m/s}$$

$$I_A^{-1} = \frac{79 \times 2^2 \times 10^{-6}}{.06^2} \left[1 - ½ (\quad) \right]$$

$$= .0878 \quad 1 - \frac{.0878}{2} \quad = .0878 \times .9561 = .0839$$

I_A = 10.8 dB inadequate

Double Canceller

$$I_A^{-1} = \frac{1.25 \times 10^4 \times 2^4 \times 10^{-12}}{.06^4} + \frac{105 \times 2^2 \times (5 \times 10^{-5})^2}{.06^2}$$

$$= .0154 + .0003 = .0157$$

I_A = 18.0 dB OK

∴ A double canceller with 10% stagger is adequate for both the land clutter and rain.

References

[1] M. I. Skolnik, *Introduction to Radar Systems*, McGraw-Hill, 1962.

[2] D. K. Barton, *Radar System Analysis*, Prentice-Hall, 1964.

[3] F. Nathanson, *Radar Design Principles*, McGraw-Hill, 1969.

[4] M. I. Skolnik, *Radar Handbook*, McGraw-Hill, 1970.

[5] B. D. Steinberg, "MTI Radar Filters" in *Modern Radar*, (R. S. Berkowitz, ed.), John Wiley & Sons, 1965.

[6] E.J. Barlow, "Doppler Radar," *Proceedings of IRE*, pages 340-355, April 1949.

[7] D.K. Barton, "Radar Equations for Jamming & Clutter," *IEEE Trans.* Aerospace & Electronic Systems, Vol. AES-3, No. 6, Eastcon Supplement, pages 340-355, November 1967.

[8] W. W. Shrader, "MTI Radar" in (4).

List of Symbols

A = antenna aperture

A_c = area of distributed clutter resolution cell — meters^2

$A_{effective}$ = effective antenna aperture

C_i/C_o = clutter attenuation factor due to MTI processing

$C(\omega)$ = clutter power spectrum

G_{SL} = antenna pattern sidelobe level illuminating clutter cell (less than unity)

$|H(\omega)|^2$ = MTI delay line canceller power transfer function

$|H_n(\omega)|^2$ = power transfer function for n delay line canceller $n = 1, 2, \ldots$

I = improvement factor

$$= \frac{\text{clutter attenuation factor}}{\text{signal attenuation factor}}$$

I_{2p} = single canceller improvement factor

I_{3p} = double canceller improvement factor

$I_{A_{2p}}$ = improvement factor limitation imposed by antenna motion for single canceller

$I_{A_{3p}}$ = improvement factor limitation imposed by antenna motion for double canceller

I_R = required improvement factor

I_T = total improvement factor available from radar

I_i	= improvement factor due to i^{th} contributor
L_{BS}	= 1.45 beam shape loss
N	= noise power
R_c	= range of clutter — meters
R_m	= maximum range at which moving target is to be detected in distributed clutter — meters
S	= signal power
$(S/N)_o$	= detectability factor, the value of signal-to-noise ratio required to achieve a given probability of detection with fixed probability of false alarm
S_i/C_i	= signal-to-clutter ratio at input of MTI system
S_i/S_o	= signal attenuation factor due to MTI processing
S_o/C_o	= signal-to-clutter ratio at output of MTI system
$S(\omega)$	= signal power spectrum
T	= antenna tilt angle — degrees
V_b	= 1st blind speed — meters/second
V_{b_n}	= n^{th} blind speed — meters/second
V_c	= volume of distributed clutter resolution cell — meter^3
c	= velocity of light — meters/second
f_m	= mean frequency of clutter spectrum — hertz
f_o	= transmission frequency — hertz
f_r	= pulse repetition frequency — hertz
n_o	= number of hits in 3 dB beamwidth
t_o	= interpulse period — seconds
Δ	= fraction of interpulse period
α	= Barlow's dimensionless stability parameter
δ	= stagger interval — seconds
η	= volume clutter reflectivity — meter^2/meter^3
θ_A	= azimuth 3 dB beamwidth — radians
θ_E	= elevation 3 dB beamwidth — radians
λ_o	= wavelength of transmission — meters
ξ	= scan angle from array normal at which electronic scan beam intercepts clutter resolution cell — degrees
σ_o	= area clutter reflectivity

σ_A	= clutter spectrum broadening due to antenna motion — hertz
σ_T	= total clutter spectrum spread — hertz
σ_c	= input clutter cross section or power
σ_i	= clutter spectrum spread due to i^{th} contributor — hertz
σ_p	= standard deviation of clutter spectrum — hertz (frequency spread of clutter)
σ_t	= radar target cross section or power
σ_v	= velocity spread — meters/second
τ	= effective pulse width — seconds
ω_A	= antenna angular rotation rate — radians/second

Sine Space Co-ordinates (Array Coordinates)

Chapter Seven

7.1 Definition — The antenna patterns and scanning properties of mechanically-scanned antennas are readily described in a spherical coordinate system with the origin at the phase center of the antenna. Since the array antenna is stationary while the beam is being scanned, the beam broadens and the beam shape changes slightly as the beam is scanned off broadside (see Fig. 6.2 and p. 158). It is, therefore, desirable to select a more convenient coordinate system to describe antenna patterns and scanning properties of electronically scanned array antennas. The advantage of sine space coordinates is that the antenna pattern remains invariant with scan angle and merely undergoes a translation that is proportional to the differential phase delay between adjacent radiating elements. Sine space, or "sin θ space", is simply the projection of the unit sphere on the plane of the array (Fig. 7.1) (Ref. 1). Rather than describe beamwidth and angular positioning increments in degrees, radians or milliradians; it is convenient in array applications to describe them in terms of their sines or the increment of their sines which are invariant with scan angle. Therefore, instead of specifying beamwidth in milliradians it is specified in millisines. Intimate knowledge of sine space coordinates and conversion to other coordinate systems is indispensable to the ESRS architect. It is an essential tool for judicious use of parameter management and determination of radar system parameters as will be explained and demonstrated in this and succeeding chapters.

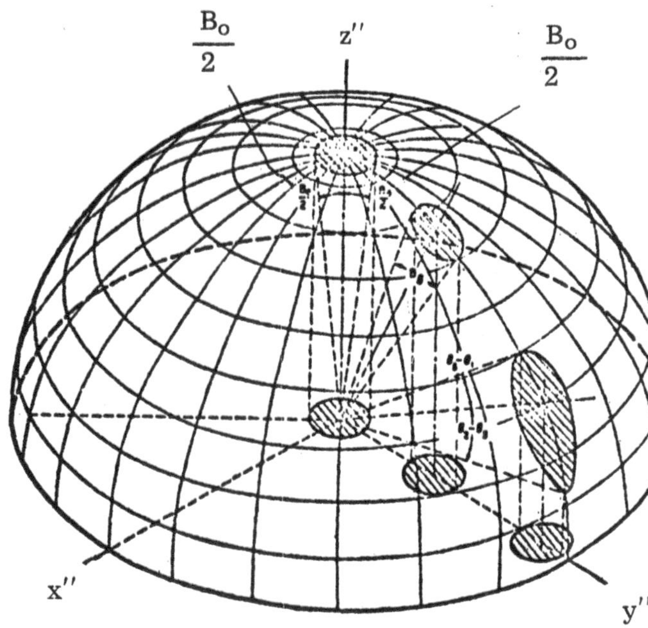

Figure 7-1

Phased Array Beam Scanning

7.2 Array-Space Coordinate Summary — To aid in the visualization of the sine space coordinate system the coordinate conversion formulas are derived by simple axes rotations of cartesian coordinates rather than the use of matrices or unit vectors. The conversion formulas are derived in Appendix C and a summary is reproduced in this section. The pertinent coordinate systems are shown and defined in Fig. 7.2. The formulas obtained are of practical design use to the ESR system engineer.

$\theta, \phi \rightarrow \alpha, \beta$

$$\alpha = \sin A = \sin\theta \cos\phi \tag{7.1}$$

$$\beta = \sin B = \sin\theta \sin\phi \tag{7.2}$$

$\alpha, \beta \rightarrow \theta, \phi$

$$\theta = \sin^{-1} [\alpha^2 + \beta^2]^{1/2} = \sin^{-1} [\sin^2 A + \sin^2 B]^{1/2} \tag{7.3}$$

$$\phi = \tan^{-1}\left(\frac{\alpha}{\beta}\right) = \tan^{-1}\left(\frac{\sin A}{\sin B}\right) \tag{7.4}$$

$A_z, E \rightarrow \alpha, \beta$

$$\alpha = \cos E \sin A_z \cos N_t - \cos E \cos A_z \sin N_t \tag{7.5}$$

$$\beta = -\cos E \sin A_z \sin N_t \sin T + \sin E \cos T - \cos E \cos A_z \cos N_t \sin T \tag{7.6}$$

$\alpha, \beta \rightarrow A_z, E$

$$E = \sin^{-1} [\beta \cos T + \sqrt{1-\alpha^2-\beta^2} \sin T] \tag{7.7}$$

$$A_z = \tan^{-1}\left[\frac{\alpha \cos N_t - \beta \sin N_t \sin T + \sqrt{1-\alpha^2-\beta^2} \sin N_t \cos T}{-\alpha \sin N_t - \beta \cos N_t \sin T + \sqrt{1-\alpha^2-\beta^2} \cos N_t \cos T}\right] \tag{7.8}$$

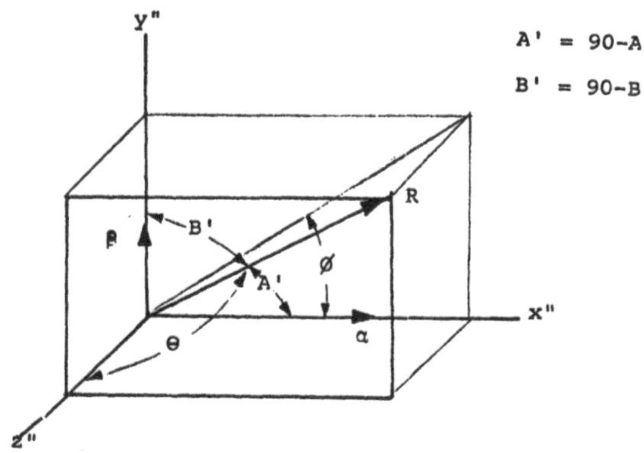

Spherical & Array Coordinates

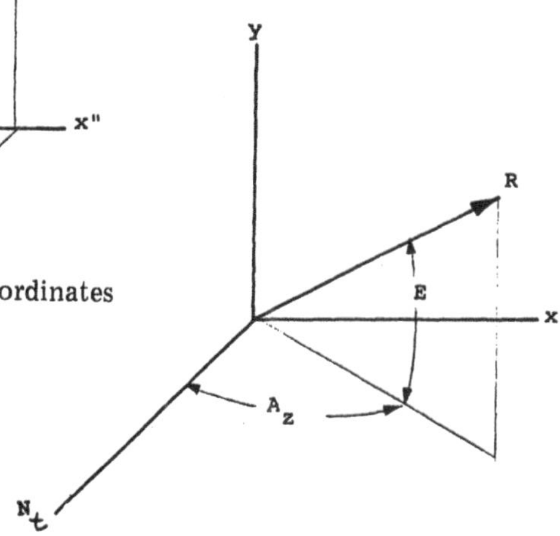

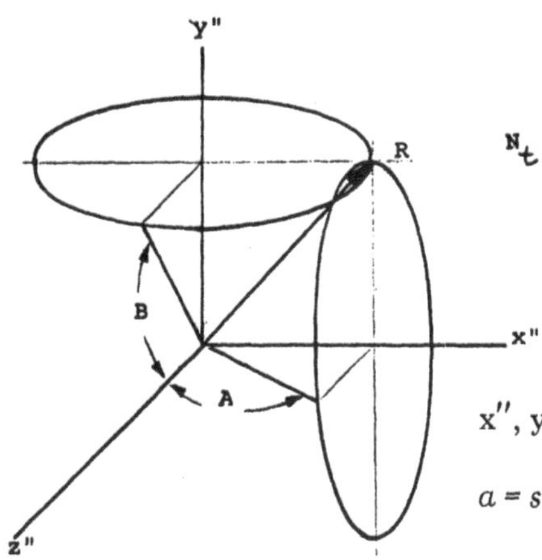

Figure 7-2

Space Array Coordinates

x'', y'' = *plane of array*

$$\alpha = \sin A = \frac{x''}{R}$$

$$\beta = \sin B = \frac{y''}{R}$$

θ, ϕ = *antenna spherical coordinates*

A_z = *azimuth angle*

E_L = *elevation angle*

T = *tilt angle of array from vertical*

N_t = *angle between true north and plane ⊥ to face of array*

x, y, z = *Cartesian coordintes in space*

$A_z, E \rightarrow \theta, \phi$

$$\theta = \cos^{-1}\left[\cos E \sin A_z \sin N_t \cos T + \sin E \sin T + \cos E \cos A_z \cos N_t \cos T\right] \quad (7.9)$$

$$\phi = \tan^{-1}\left[\frac{-\cos E \sin A_z \sin N_t \sin T + \sin E \cos T - \cos E \cos A_z \cos N_t \sin T}{\cos E \sin A_z \cos N_t - \cos E \cos A_z \sin N_t}\right] \quad (7.10)$$

$\theta, \phi \rightarrow A_z, E$

$$E = \sin^{-1}\left[\sin\theta \sin\phi \cos T + \cos\theta \sin T\right] \quad (7.11)$$

$$A_z = \tan^{-1}\left[\frac{\sin\theta \cos\phi \cos N_t - \sin\theta \sin\phi \sin N_t \sin T + \cos\theta \sin N_t \cos T}{-\sin\theta \cos\phi \sin N_t - \sin\theta \sin\phi \cos N_t \sin T + \cos\theta \cos N_t \cos T}\right] \quad (7.12)$$

$R, A_z, E \rightarrow x, y, z$

$$x = R \cos E \sin A_z \quad (7.13)$$

$$y = R \sin E \quad (7.14)$$

$$z = R \cos E \cos A_z \quad (7.15)$$

$x, y, z \rightarrow R, A_z, E$

$$R = \left[x^2 + y^2 + z^2\right]^{1/2} \quad (7.16)$$

$$E = \sin^{-1}\left(\frac{y}{R}\right) \tag{7.17}$$

$$A_z = \tan^{-1}\left(\frac{x}{z}\right) \tag{7.18}$$

$R, \theta, \phi \rightarrow x'', y'', z''$

$$x'' = R\sin\theta\cos\phi \tag{7.19}$$

$$y'' = R\sin\theta\sin\phi \tag{7.20}$$

$$z'' = R\cos\theta \tag{7.21}$$

$x'', y'', z'' \rightarrow R, \theta, \phi$

$$R = [x''^2 + y''^2 + z''^2]^{1/2} \tag{7.22}$$

$$\phi = \tan^{-1}\left(\frac{y''}{x''}\right) \tag{7.23}$$

$$\theta = \cos^{-1}\left(\frac{z''}{R}\right) \tag{7.24}$$

$x'', y'', z'' \rightarrow x, y, z$

$$x = x'' \cos N_t - y'' \sin N_t \sin T + z'' \sin N_t \cos T \tag{7.25}$$

$$y = y'' \cos T + z'' \sin T \tag{7.26}$$

$$z = -x'' \sin N_t - y'' \cos N_t \sin T + z'' \cos N_t \cos T \tag{7.27}$$

$x, y, z \rightarrow x'', y'', z''$

$$x'' = x \cos N_t - z \sin N_t \tag{7.28}$$

$$y'' = -x \sin N_t \sin T + y \cos T - z \cos N_t \sin T \tag{7.29}$$

$$z'' = x \sin N_t \cos T + y \sin T + z \cos N_t \cos T \tag{7.30}$$

7.3 Element Spacing and Grating Lobe Location — If the element spacing exceeds half a wavelength in a phased array, additional lobes (known as grating lobes) can appear in the antenna radiation pattern with amplitudes equal to that of the main beam. They are due to the radiation from the elements adding in phase and forming additional wavefronts in space directions for which the relative path lengths are integral multiples of 2π.

Grating lobes occur when

$$\psi_\gamma \leqslant \psi_\theta + 2\pi \tag{7.31}$$

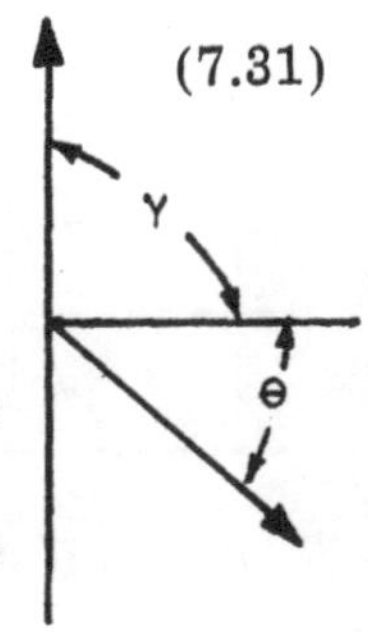

where

ψ_θ = differential phase shift between adjacent elements to form beam in θ direction

ψ_γ = differential phase shift between adjacent elements that forms grating lobe in γ direction

From (3.3), the condition for suppression of a grating lobe is

$$\frac{2\pi d}{\lambda} \sin\gamma \leqslant \frac{2\pi d}{\lambda} \sin(-\theta) + 2\pi$$

or

$$\frac{d}{\lambda} (\sin\gamma + \sin\theta) \leqslant 1$$

$$d \leqslant \frac{\lambda}{\sin\gamma + \sin\theta} \tag{7.32}$$

In a practical system the element spacing is chosen to prevent the grating lobes from appearing at the end of the array ($\gamma = 90°$), at the highest operating frequency of the radar, for the maximum scan angle.

$$\therefore d = \frac{\lambda}{1 + \sin\theta_m} \tag{7.33}$$

where

d = interelement spacing

λ = transmission wavelength

θ_m = maximum scan angle from array normal

In determining the element spacing the array designer uses a value θ_m of 5 to 10 degrees larger than the maximum scan angle to avoid the severe element mismatch that occurs near the scan angle and produces a grating lobe.

An array is designed to avoid grating lobes because they lead to ambiguities and lower directive antenna gain. Targets viewed by the grating lobes cannot be distinguished from targets viewed by the main beam. If the grating lobe is equal in amplitude to the main beam, the antenna gain is the same in both directions and consequently is 3 dB less in the main beam than it would be in the absence of grating lobes.

The ESR system designers usually attempt to space the array elements as far apart as is consistent with angular coverage requirements in order to reduce the number of radiating elements to a minimum for a specified beamwidth. Fewer elements mean fewer phase shifters, phase shifter drivers and power supplies, and less drive power, cabling, beam steering unit complexity, cooling, maintenance, and initial and operating costs. The narrower the beamwidth, the more substantial are the savings.

Additional element savings can be achieved by judicious selection of element configuration. Rectangular (square) and triangular configurations are shown in Fig. 7.3. For unequal coverage in azimuth and elevation, rectangular or isosceles triangle element spacing is employed.

Equilateral triangle spacing for conical coverage uses the minimum number of elements. For 90° conical coverage 13% fewer elements are required than for rectangular spacing. The element impedance match is almost as good as rectangular or isosceles element spacing. The equilateral triangle spacing obtains the same loss over the total scan volume as rectangular or isosceles element spacing. For unequal coverage in the two orthogonal angular coordinates, the isosceles triangle configuration uses fewer elements than the rectangular configuration.

The procedures for locating grating lobes and their locations in θ, ϕ, spherical coordinates are shown in Figs. 7.4 and 7.5 for equilateral triangle and rectangular (square) element spacing respectively. Both configurations are designed to provide at least 90° conical coverage

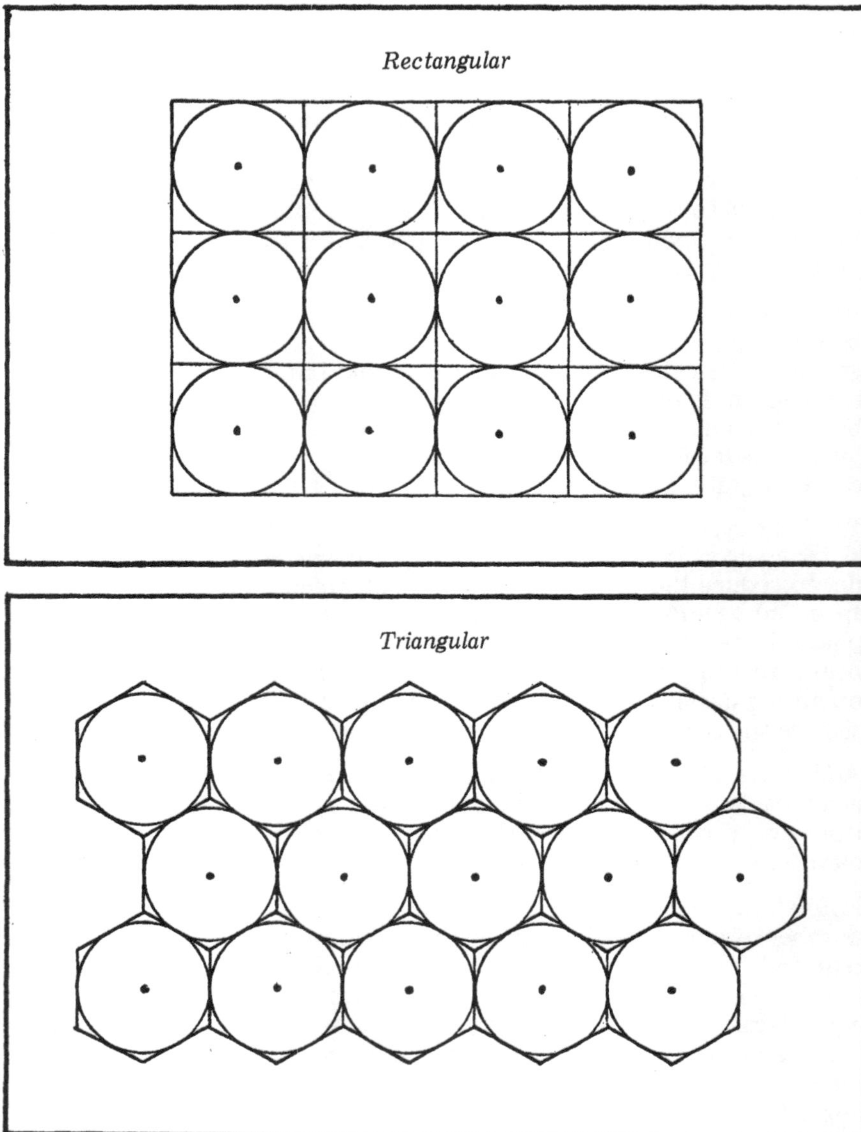

Figure 7-3

Element Configurations

($\theta_{min} = 45°$). The upper diagrams in Figs. 7.4 and 7.5 illustrate how to determine the effective element spacing at any angle ϕ, i.e., it is the projection of adjacent element positions onto a line pointing in the ϕ direction and originating on one of the elements. For the equilateral triangle configuration the grating lobes first appear in the six directions perpendicular to the sides of the triangle. For the square configuration the grating lobes first appear in the four directions perpendicular to the sides of the square.

7.4 Differential Phase Shift Between Elements — The differential phase shift required between elements to scan a beam an angle θ from broadside in a linear array is given by the interferometer equation:

$$\Delta\phi = \frac{2\pi d}{\lambda} \sin\theta \tag{3.3}$$

For a two dimensional array steering in a coordinate system referenced to the plane of the array the differential phase shift requirements are:

$$\Delta\phi_{x''} = \frac{2\pi dx''}{\lambda} \sin A \tag{7.34}$$

and

$$\Delta\phi_{y''} = \frac{2\pi dy''}{\lambda} \sin B \tag{7.35}$$

where

dx'' = interelement spacing in the x'' direction

dy'' = interelement spacing in the y'' direction

A&B = are defined in Fig. 7.2

Substituting for sinA and sinB from Eqs. (7.1) and (7.2), one obtains:

$$\Delta\phi_{x''} = \frac{2\pi dx''}{\lambda} \alpha = K_{x''}\alpha \tag{7.36}$$

$$\Delta\phi_{y''} = \frac{2\pi dy''}{\lambda} \beta = K_{y''}\beta \tag{7.37}$$

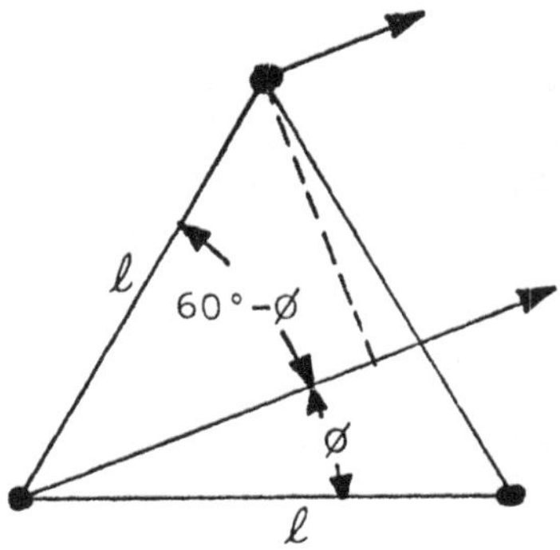

$$\sin\theta = \frac{\lambda}{d} - 1$$

$$d = \ell \cos(60^\circ - \phi) \quad 0 \leqslant \phi \leqslant 30^\circ$$

$$\therefore \sin\theta = \frac{\lambda}{\ell \cos(60^\circ - \phi)} - 1 \qquad (7.38)$$

$$\theta_{min} \; @ \; \phi = 30^\circ$$

$$\therefore \sin\theta_{min} = \frac{\lambda}{.866\ell} - 1 \quad \text{or } \ell = \frac{1.1547\,\lambda}{1 + \sin\theta_{min}} - 1$$

$$\therefore \ell = .6764\lambda \text{ for } \theta_{min} = 45^\circ$$

To find ϕ for $\theta_{max} = 90^\circ$ $\quad 1 = \dfrac{\lambda}{.6764\lambda\cos(60^\circ - \phi)} - 1$

Or $\quad \cos(60^\circ - \phi) = \dfrac{1}{2 \times .6764} = .7392$

$$\phi = 17.7^\circ$$

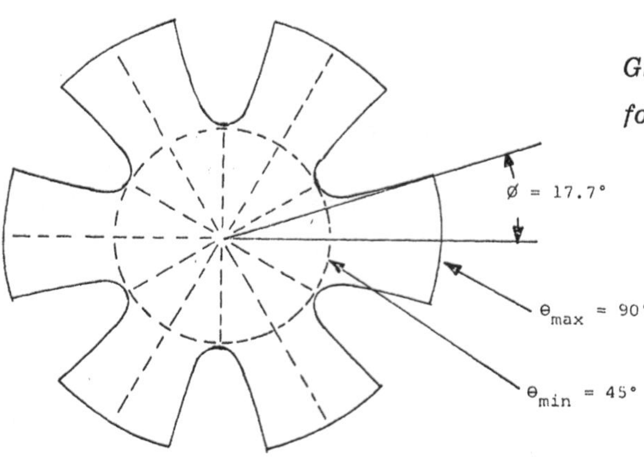

Figure 7-4

Grating Lobe Location for Triangular Spacing

$$\sin\theta = \frac{\lambda}{d} - 1$$

$$d = \ell\cos\phi \quad 0 \leqslant \phi \leqslant 45^\circ$$

$$\therefore \sin\theta = \frac{\lambda}{\ell\cos\phi} - 1 \tag{7.39}$$

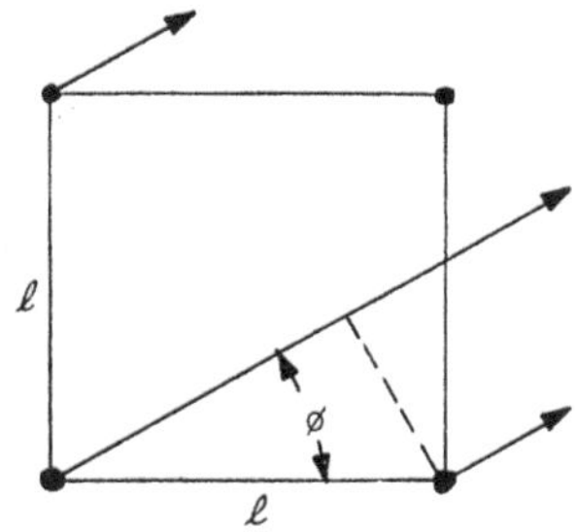

$$\theta_{min} \ @ \ \phi = 0^\circ$$

$$\therefore \sin\theta_{min} = \frac{\lambda}{\ell} - 1 \qquad \text{or} \quad \ell = \frac{\lambda}{1 + \sin\theta_{min}}$$

$$\ell = .586\lambda \ \text{ for } \theta_{min} = 45^\circ$$

To find ϕ for $\theta_{max} = 45^\circ$ $\quad 1 = \dfrac{\lambda}{.586\lambda\cos\phi} - 1$

$$\text{or } \cos\phi = \frac{1}{2 \text{ x } .586} = .8536$$

$$\phi = 31.4^\circ$$

Figure 7-5
Grating Lobe Location for Rectangular Spacing

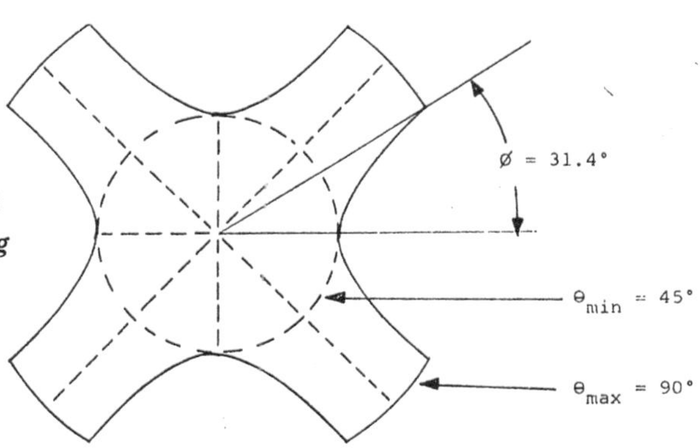

These equations indicate that the interelement differential phase shifts required to steer the beam are proportional to the sine space coordinates α, β. In an M by N array, the phase steering command at each element is

$$\phi_{mn} = m\Delta\phi_{x''} + n\Delta\phi_{y''} \qquad (\text{Modulo } 2\pi)$$

where

m	= 1, 2 . . . M	Row
n	= 1, 2 . . . N	Column

In most phased array designs row and column steering is employed. This simplifies the beam steering unit computations and reduces the number of steering command cables from MN to M + N. Row and column steering requires the addition of an adder register with modulo 2π capability at each phase shifter.

For equilateral triangle element spaced arrays, the use of phase gradients, γ, δ, normal to two sides of the triangle as shown below, minimizes the total number of rows and columns required to drive the array and further simplifies computations of the beam steering unit.

The γ, δ steering commands are related to the α, β sine space coordinates as shown.

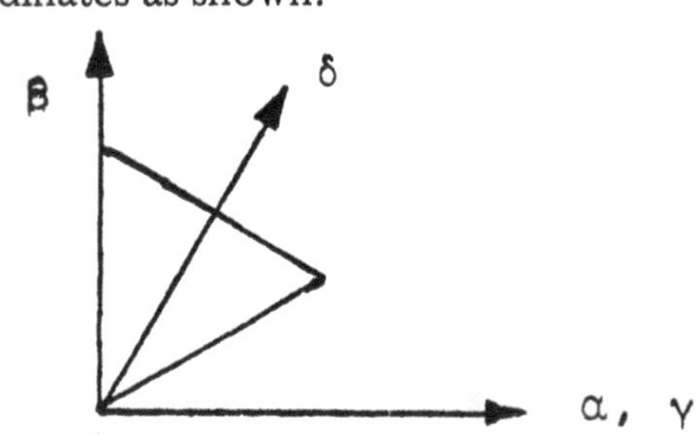

$$\gamma = \alpha - \beta \qquad (7.40)$$

$$\delta = 2\beta \qquad (7.41)$$

7.5 Beamwidth — The angular resolution and accuracy of a radar are dependent on the beamwidth. In a phased array radar the beamwidth is a function of scan angle from array normal. Consider a circular beam at array normal as shown in Fig. 7.1.

$$b = 2 \sin \frac{B_o}{2} \approx \sin B_o \qquad (7.42)$$

where

B_o = beamwidth at array normal (zero scan angle)

b = beamwidth in sine space (at any scan angle)

When the beam is directed away from the array normal then the beamwidth, B_{θ_1}, at scan angle θ_1 is

$$B_{\theta_1} = \theta_{1+} - \theta_{1-} \tag{7.43}$$

where

$$\sin\theta_{1+} = \sin\theta_1 + \frac{b}{2} = \sin\theta_1 + \sin\frac{B_o}{2} \tag{7.44}$$

$$\sin\theta_{1-} = \sin\theta_1 - \frac{b}{2} = \sin\theta_1 - \sin\frac{B_o}{2} \tag{7.45}$$

or

$$B_{\theta_1} = \sin^{-1}\left[\sin\theta_1 + \sin\frac{B_o}{2}\right] - \sin^{-1}\left[\sin\theta_1 - \sin\frac{B_o}{2}\right] \tag{7.46}$$

A simpler expression may be obtained by noting from Eqs. (7.44) and (7.45)

$$b = \sin\theta_{1+} - \sin\theta_{1-} \tag{7.47}$$

$$b = 2\sin\tfrac{1}{2}(\theta_{1+} - \theta_{1-}) \cos\tfrac{1}{2}(\theta_{1+} + \theta_{1-}) \tag{7.48}$$

or

$$\sin\tfrac{1}{2}(\theta_{1+} - \theta_{1-}) = \frac{\frac{b}{2}}{\cos\tfrac{1}{2}(\theta_{1+} + \theta_{1-})} \tag{7.49}$$

for narrow beams and $\theta_1 \leqslant 60°$ (7.50)

$$B_o = b$$

and

$$\theta_1 = \tfrac{1}{2}(\theta_{1+} + \theta_{1-})$$

$$\therefore \quad B_{\theta_1} = \frac{b}{\cos\tfrac{1}{2}(\theta_{1+} + \theta_{1-})} \approx \frac{B_o}{\cos\theta_1} = B_o \sec\theta_1 \qquad (7.51)$$

Expression (7.51) indicates the beam broadens in the θ direction (refer to Fig. 7.1) as the secant of the scan angle, θ_1.

7.6 Tilt Angle — The tilt angle is defined as the angle the array face is inclined from the vertical. The selection of tilt angle for a particular ESR application involves any one or a combination of the following four considerations.

1. Angular Coverage Equalization — The number of elements in a phased array is minimized and the beam steering unit design is simplified by selecting the tilt angle to equalize the scan to the extreme points of coverage. Fig. 7.6 illustrates how the tilt angle is determined for hemispherical coverage when using 3 or 4 phased array radars. The results are summarized below.

No. of Radars	Max Azimuth Coverage	Max Scan Angle	Tilt Angle
4	±45°	54.7°	35.3°
3	±60°	63.5°	26.5°

In the many applications where the coverage is defined as a segment of a hemisphere it is convenient to determine the tilt angle that minimizes the scan angle by use of Meridional Stereographic Projection Charts. These charts are obtainable from the U.S. Navy Hydrographic Office in Washington, D.C.

An illustrative example of the use of Meridional Stereographic chart is given in Fig. 7.7 for a phased array coverage requirement of ±40° in azimuth and 0° - 50° in elevation. The procedure is as follows:

Step 1: Draw two lines OA and OD from the center of the graph to the $-40°$ and $+40°$ points respectively of the outer circle. AOD represents the projection of $\pm 40°$ azimuth and $0° - 90°$ elevation coverage surface of a hemisphere onto a circle when viewed from the top of the sphere.

MAKE SCAN TO EXTREME POINTS EQUAL

$\angle J = \angle H$

SCAN TO ZENITH

$\angle F + \angle J = 90°$

$\angle F + \angle H = 90°$

FOR RIGHT SPHERICAL TRIANGLE

$\cos H = \cos F \cos G$

OR

$\cos H = \cos (90° - H) \cos G$

$$\therefore \tan H = \sec G \tag{7.52}$$

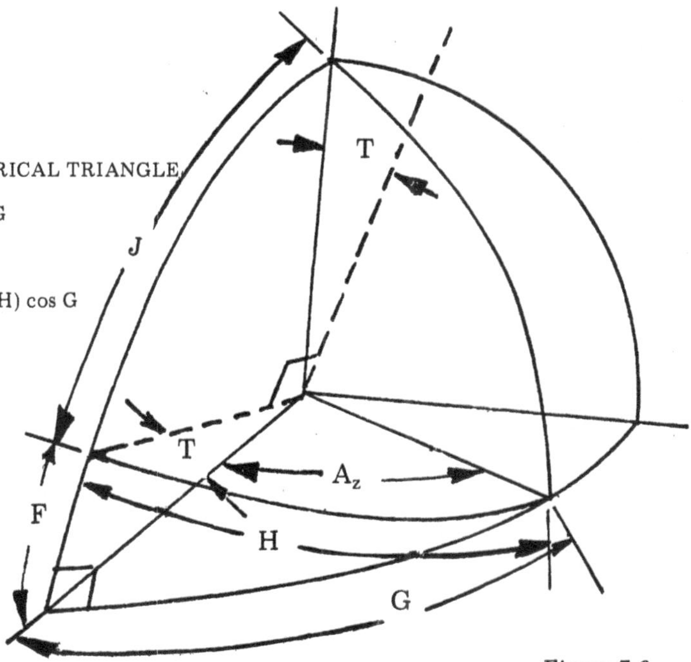

<u>4 RADARS</u>

$A_z = G = 45°$

$\tan H = \sec 45° = 1.414$

$H = 54.7°$ MAXIMUM SCAN ANGLE

$F = T = 35.3°$ TILT ANGLE

<u>3 RADARS</u>

$A_z = G = 60°$

$\tan H = \sec 60° = 2.0$

$H = 63.5°$ MAXIMUM SCAN ANGLE

$F = T = 26.5°$ TILT ANGLE

Figure 7-6
Tilt Angle for Hemispherical Coverage

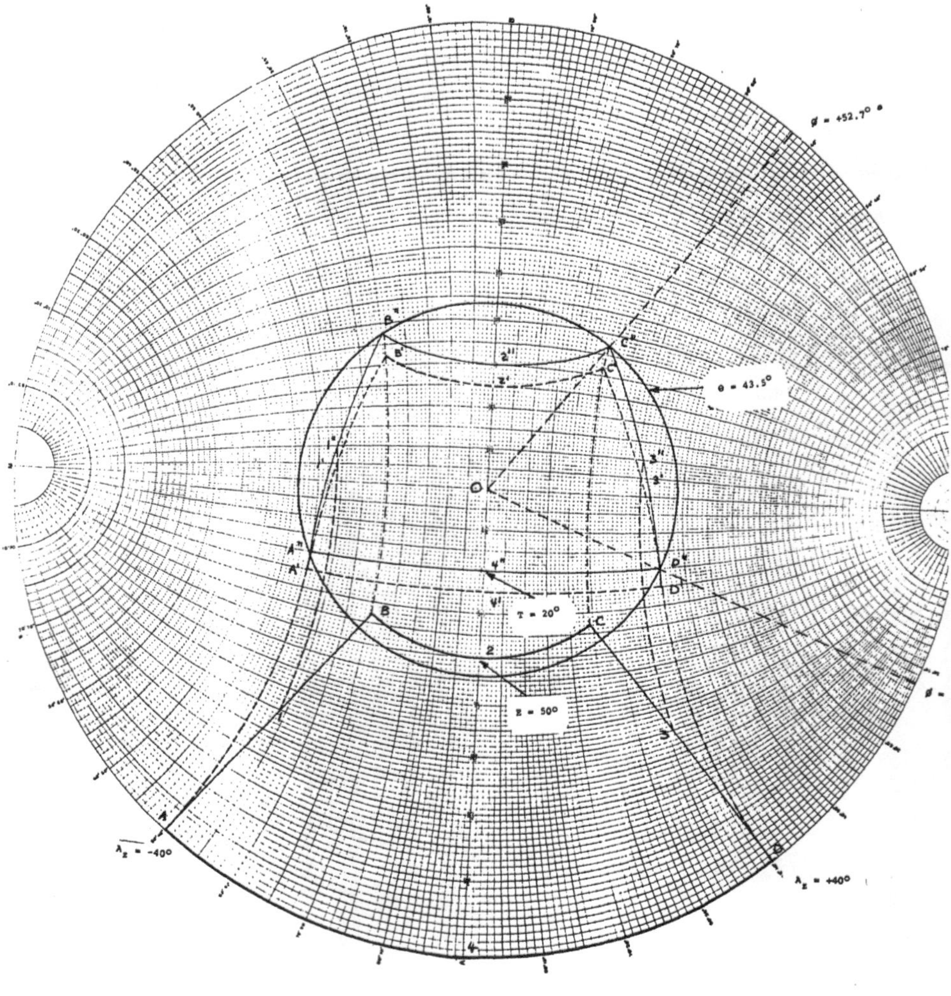

Figure 7-7

— Tilt & Scan Angle Determination
Azimuth Coverage ± 40°

— Elevation Coverage 0° — 50°

Step 2: Draw an arc of a circle with radius O2 until it intersects AOD at B and C where O2 is equal to 40°(90° – elevation coverage). Now ABCD represents the projection of ±40° azimuth and 0° - 50° elevation coverage surface of a hemisphere onto a circle when viewed from the top of the sphere.

Step 3: As a first iteration translate ABCD to A′B′C′D′ along isoangle lines as shown. The translation is 65° [90° – ½ (elevation coverage)]. The tilt angle is 04′ or 25°. This tilt angle equalizes the scan angle coverage in elevation (±25°) at zero azimuth angle but does not equalize the scan angle coverage to the extreme points (OC′ ≠ OD′ and OB′ ≠ OA′).

Step 4: On the second iteration with the aid of a compass it is readily determined that an additional 5° translation to A″B″C″D″ equalized the scan angle coverage to the extreme points (OC″ = OD″ = OB″ = OA″ = θ_{max} = 43.5°). The required tilt angle is 04″ or 20°. The corresponding ϕ values may be obtained by extending lines OC″, OD″, OB″ and OA″ to the outer circle.

A more exact determination of tilt angle can be obtained by a graphical solution (or iteration on a calculator) of Eq. (7.9) for the two extreme scan points (E = 0°, A_z = 40° and E = 50°, A_z = 40°).

$$\theta = \cos^{-1}[\cos 40^\circ \cos T] = \cos^{-1}[\sin 50^\circ \sin T + \cos 50^\circ \cos 40^\circ \cos T]$$

or

$$\cos^{-1}[.7660 \cos T] = \cos^{-1}[.7660 \sin T + .4924 \cos T]$$

A comparison of the results obtained from the Meridional Stereographic Projection chart and Eqs. (7.9) and (7.10) is shown in the following table. The values in parenthesis are obtained from the equations.

T	θ	ϕ	A_z	E
20° (19.7°)	43.5° (43.8°)	52.7° (53.3°)	40°	0
		–22.2°(-21.9°)	40°	50°

2. Maximum Array Gain — It is often desirable to point the array normal in the direction where the targets will be viewed at maximum range. For aircraft targets, maximum range detection is on the horizon, i.e., zero tilt angle. The tilt angle design may also be a compromise between providing maximum gain in the direction of maximum detection range and minimum scan angle.

If the ESR is being used to monitor reentry vehicles, the tilt angle is made equal to the average reentry angle; possibly modified to assure adequate coverage without an excessive scan angle design.

3. Beamwidth Management — The beam broadening characteristic of a phased array as a function of scan angle from the array normal can be employed advantageously to minimize the number of beam positions required for elevation coverage. For example, 0° to 45° elevation coverage with a 1° beamwidth requires 44 elevation beam positions (spaced a beamwidth apart) for a tilt angle of +22.5°, and only 40 elevation beam positions for a tilt angle of 0° (refer to section 7.8). The 0° tilt angle requires a larger scan angle capability than the +22.5° tilt angle but it has the advantage of providing narrow high gain beams in the maximum range direction for aircraft detection and wide low gain beams, at the high elevation angles where altitude limited aircraft are at short range so that neither the resolution nor the gain are required.

4. Maximize Coverage — The tilt angle may be used to move the grating lobes away from a coverage sector. For example, if a phased array is designed with equilateral triangle element spacing to provide 45° scan capability and it is desired to include a pencil beam raster search on the horizon to ±60° in azimuth, this can be accomplished by using 0° tilt angle (see Fig. 7.4).

7.7 Array Design

Element Area

1. Rectangular element spacing (see Fig. 7.3)

$$a_R = \ell_x'' \ell_y'' \tag{7.53}$$

and

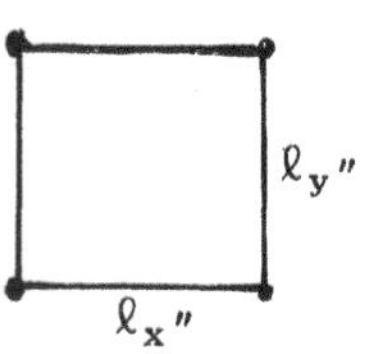

$$d_e = \begin{cases} \ell_x'' & (\ell_x'' > \ell_y'') \\ \text{or} & \\ \ell_y'' & (\ell_y'' > \ell_x'') \end{cases} \tag{7.54}$$

$$\therefore a_R = d_e^2 \qquad (7.55)$$

where:

- a_R = element area for rectangular element spacing
- $\ell_{x''}$ = element spacing along x″ coordinate
- $\ell_{y''}$ = element spacing along y″ coordinate
- d_e = effective element spacing in direction where grating lobes first appear

2. *Equilateral triangle element spacing* (see Fig. 7.3)

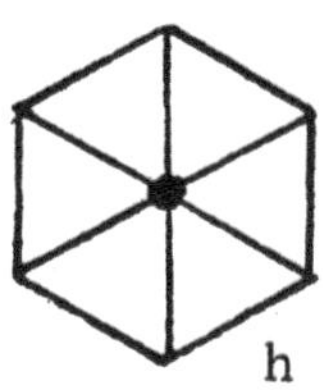

$$a_\Delta = 6 \times 1/2 \times h \times .866h = 2.6h^2 \qquad (7.56)$$

$$\tan 30° = \frac{h/2}{\ell/2}$$

or

$$h = .5774\ell \qquad (7.57)$$

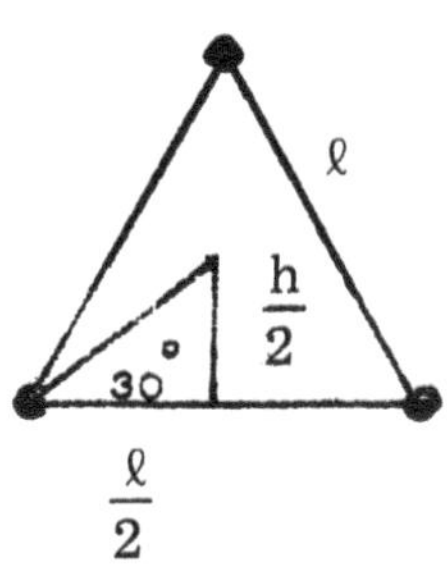

and

$$a_\Delta = .866\ell^2 \qquad (7.58)$$

$$d_e = .866\ell \quad \text{(see Fig. 7.4)} \qquad (7.59)$$

$$\therefore a_\Delta = 1.1547d_e^2 \qquad (7.60)$$

where:

- a_Δ = element area for equilateral element spacing
- h = side of circumscribed hexagon (see Fig. 7.3)
- ℓ = element spacing for equilateral triangle configuration
- d_e = effective element spacing in direction where grating lobes first appear

Number of Elements

$$N_E = \frac{A}{a} \tag{7.61}$$

where:

N_E = number of elements

A = aperture size

a = element area

The aperture size is dependent on the beamwidth, illumination taper, transmission wavelength and shape of the array.

$$\theta_o = \frac{K_o \lambda}{L} \tag{7.62}$$

where:

θ_o = beamwidth in one dimension — degrees

K_o = 50.4 to 74 (depending on illumination taper) many existing arrays are between 70 to 72 for *measured* maximum sidelobe levels between 25 to 28 dB

λ = transmission wavelength — meters

L = length of linear array — meters

The table below contains the most common phased array shapes and the corresponding array area.

Array Shape	Array Area
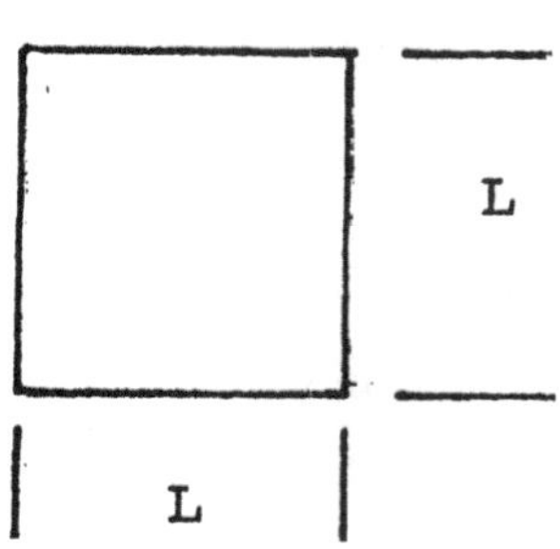	L^2

(continued)

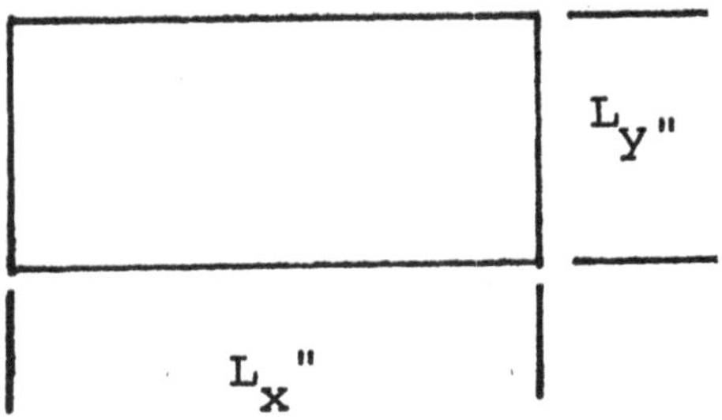

$L_{x''} L_{y''}$

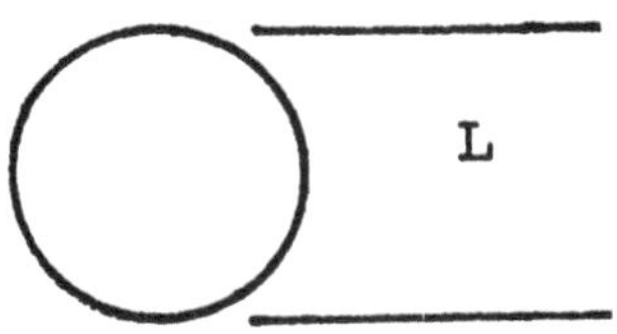

$.7854\ L^2$

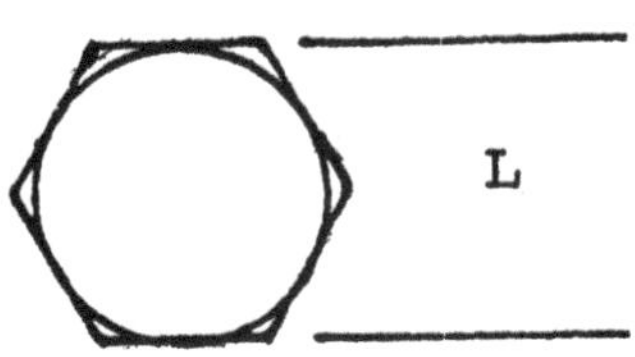

$.866\ L^2$

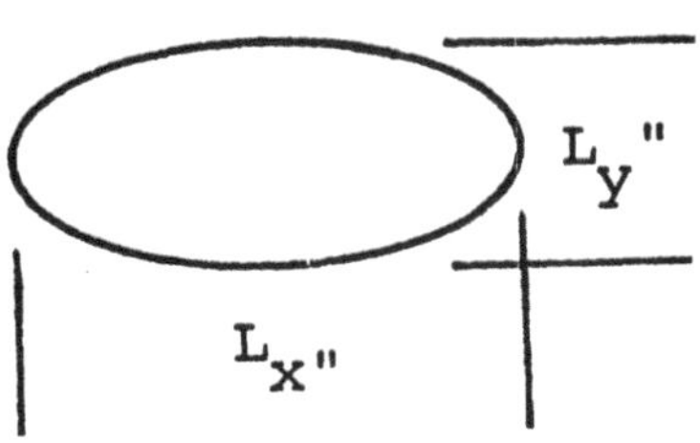

$.7854\ L_{x''} L_{y''}$

7.8 Search Pattern — The search sector is scanned with columned or interlaced beams. For a circular beam the interlaced pattern requires 13.4% (1 - .866) more elevation beam positions compared to columned beam but it is still preferred and more commonly employed because the interlaced pattern decreases the depths of the holes between beam positions.

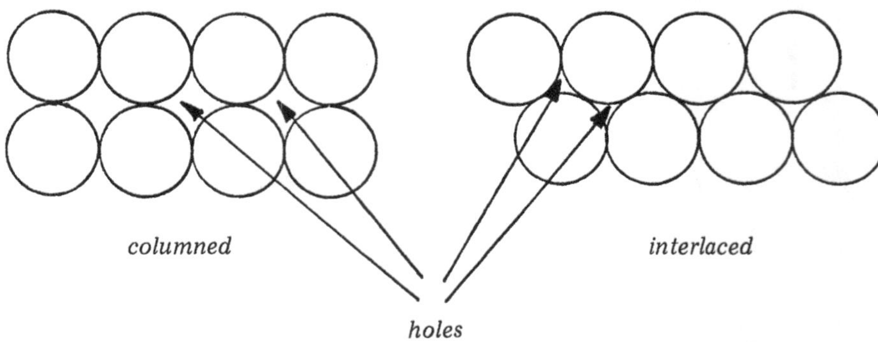

A second consideration is the size of the beam position steps. The pattern loss (see Refs. 2 and 3) decreases as the beam stepping increment decreases but more beam positions and, therefore, more time are required to cover the search sector. The most common preference is the selection of the 3 dB beamwidth steps. However, there are situations where a very high probability of detection is required and smaller increments are employed (see Refs. 2, 3, and 4), especially if the coverage sector is small.

Cross Over	Step Size	Pattern Loss
3.0 dB	θ_o	3.2 dB
2.2 dB	$.86\theta_o$	2.6 dB (typical)

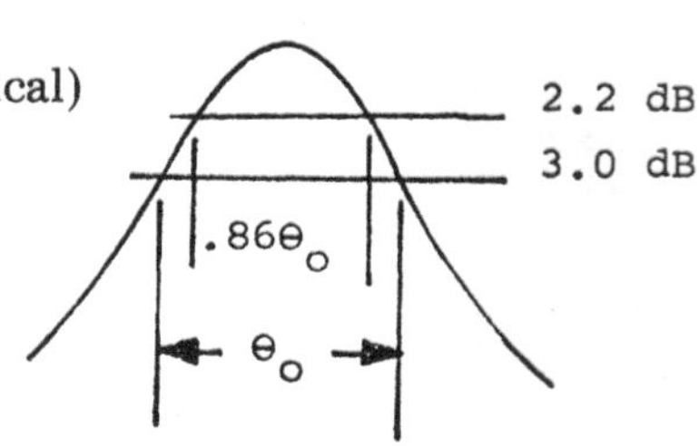

(Gaussian Pattern)

7.9 Use of Sine Space Coordinates in ESR Design

Illustrative Example — Design an array with a 1.15° (20 ms) pencil beam and maximum sidelobes of 28 dB to obtain ± 45° azimuth coverage with the array normal pointing at true North, and 0° to 80° elevation coverage. The maximum range and altitude of the coverage volume are 200 nm and 100 kft respectively. The transmission frequency is 3 GHz.

Determine:
1. tilt angle and maximum scan angle
2. sine space coverage and plot on α, β coordinates
3. search frame time
4. aperture size
5. aperture gain
6. number of elements

Part 1 — *Tilt Angle & Maximum Scan Angle*

For equal coverage to extreme points ($E = 0°$, $A_z = 45°$ and $E = 80°$, $A_z = 45°$) iterating equation (7.9) obtains:

$$\theta = \cos^{-1}(\cos 45° \cos T) = \cos^{-1}[\sin 80° \sin T + \cos 80° \cos 45° \cos T]$$

or

$$\theta = \cos^{-1}(.7071 \cos T) = \cos^{-1}[.9848 \sin T + .1228 \cos T]$$

$\therefore T = 30.7°$ tilt angle

$\theta = 52.3°$ maximum scan angle

Part 2 — *Sine Space Coverage*

From Eqs. (7.5) and (7.6) with $N_t = 0$, $T = 30°$

$$\alpha = \cos E \sin A_z$$

$$\beta = \cos 30.3° \sin E - \sin 30.3° \cos E \cos A_z$$

$$\beta = .8634 \sin E - .5045 \cos E \cos A_z$$

For lower extreme $E = 0°$, $A_z = 45°$

$$\alpha_L = \sin 45° = .7071$$

$$A_L = \sin^{-1} \alpha_L = 45°$$

$$\beta_L = -.5045 \cos 45° = -.3567$$

$$\beta_L = \sin^{-1} \beta_L = -20.9°$$

For upper extreme $E = 80°$ $A_z = 45°$

$\alpha_U = \cos 80° \sin 45° = .1228$

$A_U = \sin^{-1} \alpha_U = 7.1°$

$\beta_U = .8634 \sin 80° - .5045 \cos 80° \cos 45°$

$\beta_U = .8503 - .0619 = .7883$

$B_L = \sin^{-1} \beta_U = 52.0°$

At center $A_z = 0$, $E = 0$ and $A_z = 0$, $E = 80°$

$\alpha_{CL} = 0$

$A_{CL} = 0°$

$\beta_{CU} = -.5045$

$\beta_{CL} = \sin^{-1} \beta_{CL} = -30.3°$

$\alpha_{CU} = 0$

$A_{CU} = 0$

$\beta_{CU} = .8634 \sin 80° - .5045 \cos 80°$

$\beta_{CU} = .8503 - .0876 = .7627$

$B_{CU} = -\sin^{-1} \beta_{CU} = 49.7°$

At $E = 40°$, $A_z = \pm 45°$

$\alpha_M = \pm \cos 40° \sin 45° = \pm .5417$

$A_M = \pm 32.8°$

$\beta_M = .8634 \sin 40° - .5045 \cos 40° \cos 45°$

$\beta_M = .5550 - .2733 = .2817$

$B_M = 16.4°$

These results are plotted in Fig. 7.8.

Part 3 — *Search Frame Time.* The search frame time is determined by summing the dwell times of all beam positions in the coverage volume.

The pencil beam will be scanned in 3 dB beamwidth increments and interlaced in elevation. (Compromise between pattern loss and frame time.)

At each elevation beam position the elevation angle determined by the lower 3 dB point of the beam will be employed to determine the maximum range (if $R_{max} < 200$ n mi) as limited by the 100 kft altitude boundary from an elevation angle-range-altitude chart (not shown). The range used in the calculation of dwell time will be the one obtained from the chart or 200 n mi, whichever is less.

The computation and retrace time added to the maximum range echo time to establish the dwell time will be assumed to be 125 μs.

$\therefore$ A_z increments = 20 ms

E increments = .8660 x 20 = 17.32 ms

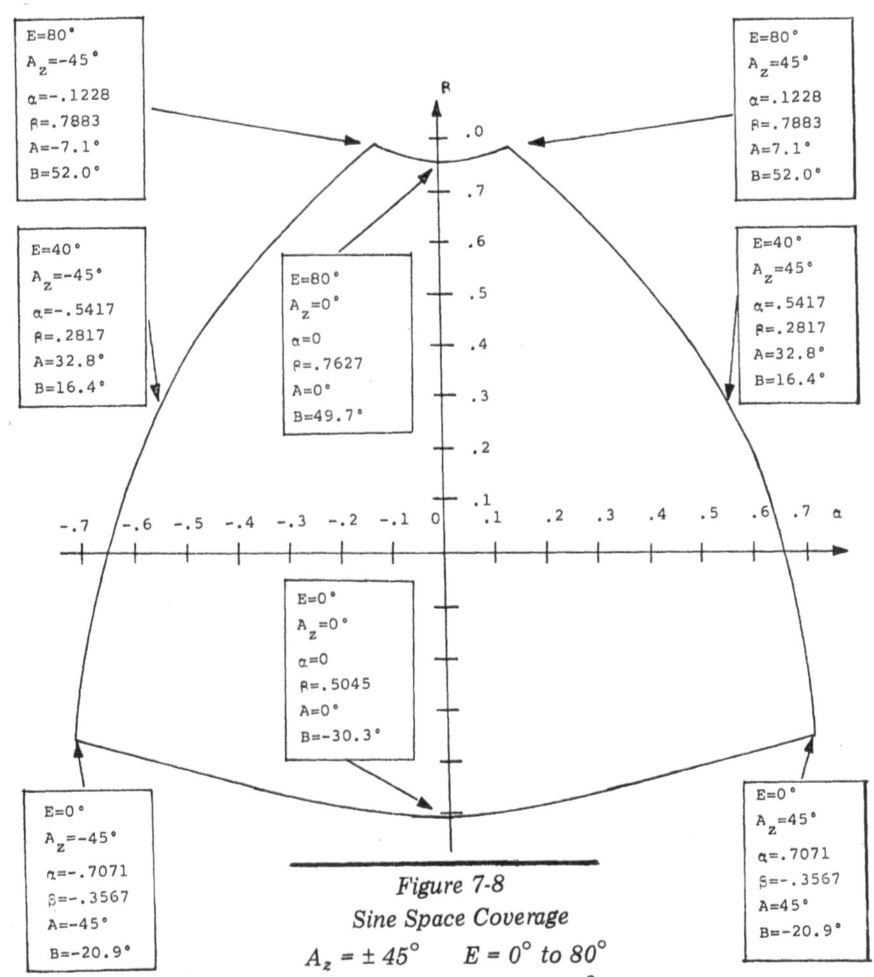

Figure 7-8
Sine Space Coverage
$A_z = \pm 45^\circ$ $E = 0^\circ$ to 80°
Tilt = 30.3° $\theta_{max} = 52.3^\circ$

Beam Position # 1

E_{L1} = 0° (lower 3 dB point of beam on horizon)

Range = 200 n m (R = 378 n m for E = 0 and 100 kft altitude)

T_1' = 200 n m x 12.36 μs/nm + 125 μs = 2597 μs

a_{M1} = cos 0° sin 45° = .7071

N_1 = Number of A_Z beam positions @ E_{L1} = $\dfrac{2a_{M1}}{.02}$

= 100 a_{M1}

N_1 = 100 x .7071 = 71

T_1 = $N_1 T_1'$ = 71 x 2597 μs = .184387 s

Beam Position # 2

E = T + B_O (where B = B_O @ A_Z = 0)

$\therefore B_O$ = E − T

sin B_{02} = sin (E_{L1} − T) + .01732

sin B_{02} = -sin 30.3° + .01732 = -.4872

B_{02} = -29.16°

E_{L2} = 30.3 − 29.16 = 1.14°

Range = 200 n m (E = 1.14° Alt. = 100 kft)

T_2' = 2597 μs

a_{M2} = cos 1.14° sin 45° = .7070

N_2 = 100 x .7070 = 71

T_2 = $N_2 T_2'$ = .184387 s

Beam Position # 36

sinB_{036} = sin(E_{L1} − T) + 35 x .01732

sinB_{036} = -sin30.3° + .6062 = .1017

B_{036} = 5.84°

E_{L36} = 30.3 + 5.84 = 36.14°

Range = 20 nm (E = 36.14° Alt = 100 kft)

T'_{36} = 20 x 12.36 + 125 = 372.2 μs

a_{M36} = cos 36.14° sin 45° = .5710

N_{36} = 100 x .5710 = 58

T_{36} = $N_{36} T'_{36}$ = 58 x 372.2 μs = .021588 s

$$\text{No. of E beam positions} = \frac{\beta_{CU} - \beta_{CL}}{.01732} = \frac{.7627 + .5045}{.01732} = 74$$

Beam Position # 74

$$\begin{aligned}
\sin B_{074} &= -\sin 30.3^\circ + 73 \times .01732 = .7598 \\
B_{074} &= 49.45^\circ \\
E_{L\,74} &= 30.3 + 49.45 = 79.75^\circ \\
\text{Range} &= 16.46 \text{ nm (100 kft)} \\
T'_{74} &= 16.46 \times 12.36 + 125 = 328.4\ \mu s \\
\alpha_{M\,74} &= \cos 79.75^\circ \sin 45^\circ = .1258 \\
N_{74} &= 100 \times .1258 = 13 \\
T_{74} &= N_{74}\ T'_{74} = 13 \times 328.4\ \mu s = .004269 \text{ s}
\end{aligned}$$

No. of E Beam Positions	$E_{L\,1}$	Range	Dwell	No. of A_z Beam Positions	T_n
	°	nml	μs		s
1	0	200	2597.0	71	.184387
2	1.14	200	2597.0	71	.184387
,					
,					
,					
,					
36	36.14	20	372.2	58	.021588
,					
,					
,					
,					
,					
,					
74	79.75	16.46	328.4	13	.004269

$$\text{Search Frame Time} = \sum_{n=1}^{74} T_n$$

Part 4 — *Aperture Size.* Measured sidelobes of 28 dB can be obtained with a Taylor illumination taper design for 30 dB sidelobes with $\bar{n} = 4$.

For this design:

$$\theta_0 = \frac{67.4\,\lambda}{L} \quad \text{CIRCULAR APERTURE}$$

[See Ref. 5, Page 256]

where $\theta_0 = 1.15°$

$$\lambda = \frac{c}{f_0} = \frac{3 \times 10^8}{3 \times 10^9} = .1 \text{ m}$$

$$L = \frac{67.4 \times .1}{1.15} = 5.86 \text{ m} \ (19.2 \text{ ft.})$$

$$\therefore \ A = .7854L^2 = .7854 \times 5.86^2$$
$$A = 27 \text{ m}^2 \ (290.4 \text{ ft.}^2)$$

Part 5 — *Aperture Gain*

From (1.10)

$$G = \frac{4\pi A}{\lambda^2} = \frac{4\pi \times 27}{0.1^2}$$

$G = 33929$

or

$G = 45.3$ dB

Part 6 — *Number of Elements*

From (7.33)

$$d_e = \frac{\lambda}{1 + \sin\theta_M} \quad \text{(where } d_e = \text{effective element spacing)}$$

Maximum scan angle = 52.3° (from Part 1)

Allowing 7.7° margin to avoid severe element mismatch near grating lobe locations,

$$\theta_M = 52.3° + 7.7° = 60°$$

$$d_e = \frac{0.1}{1 + \sin 60°} = .0536 \text{ m}$$

From (7.60)

$$a_\Delta = 1.1547\, d_e^2 \quad \text{(element area)}$$

$$a_\Delta = 1.1547 \times .0536^2 = .00332 \text{ m}^2$$

From (7.61)

$$N_E = \frac{A}{a} = \frac{27}{.00332} = 8133$$

References

[1] Wilhelm H. Von Aulock, "Properties of Phased Arrays," *Proc. IRE*, p. 1715-1727, October 1960 (Reprinted in Hansen, R.C. *Significant Phased Array Papers*, Artech House Inc., 1973)

[2] Hall, W.M. and Barton, D.K., "Antenna Pattern Loss Factor for Scanning Radars," *Proc. IEEE*, Sept. 1965, pp 1257-1258. (Reprinted in Barton, D.K., *Radars — Volume 2*, Artech House Inc., 1974).

[3] Hall, W.M., "Antenna Beam-Shape Factor in Scanning Radars," *IEEE Trans. AES-4*, No. 3, May 1968, pp 402-409. (Reprinted in Barton, D.K., *Radars — Volume 2*, Artech House Inc., 1974).

[4] Hahn, P.M. & Gross, S.D., "Beam Shape Loss and Surveillance Optimization for Pencil Beam Arrays," *IEEE Trans. on Aero Space and Electronic Systems*, pp 674-675.

[5] Barton, D.K. & Ward, H.R., *Handbook of Radar Measurement*, Prentice-Hall, 1969.

Appendix C
Array-Space Coordinate Conversion

The coordinates are defined in Figure 7.2, p. 175.

Directional Cosines

$$\alpha = \sin A = \cos A' = \frac{x''}{R} \text{ where } (A' = 90^\circ - A) \quad \text{(C-1)}$$

$$\beta = \sin B = \cos B' = \frac{y''}{R} \text{ where } (B' = 90^\circ - B) \quad \text{(C-2)}$$

$$\cos\theta = \frac{z''}{R} \quad \text{(C-3)}$$

$$x''^2 + y''^2 + z''^2 = R^2 \quad \text{(C-4)}$$

or

$$\left(\frac{x''}{R}\right)^2 + \left(\frac{y''}{R}\right)^2 = 1 - \left(\frac{z''}{R}\right)^2 \quad \text{(C-5)}$$

$$\therefore \sin^2 A + \sin^2 B = \sin^2\theta \quad \text{(C-6)}$$

or

$$\sin^2\theta = \alpha^2 + \beta^2 \quad \text{(C-7)}$$

$$\tan\phi = \frac{x''}{y''} = \frac{\alpha}{\beta} = \frac{\sin A}{\sin B} \quad \text{(C-8)}$$

N_t rotation

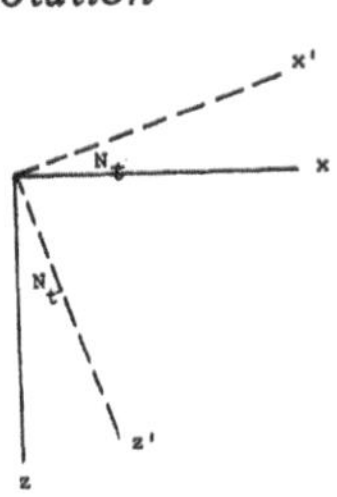

$$y' = y \quad \text{(C-9)}$$

$$x' = x\cos N_t - z\sin N_t \quad \text{(C-10)}$$

$$z' = x\sin N_t + z\cos N_t \quad \text{(C-11)}$$

or

$$y = y' \quad \text{(C-12)}$$

$$x = x'\cos N_t + z'\sin N_t \quad \text{(C-13)}$$

$$z = -x'\sin N_t + z'\cos N_t \quad \text{(C-14)}$$

T rotation

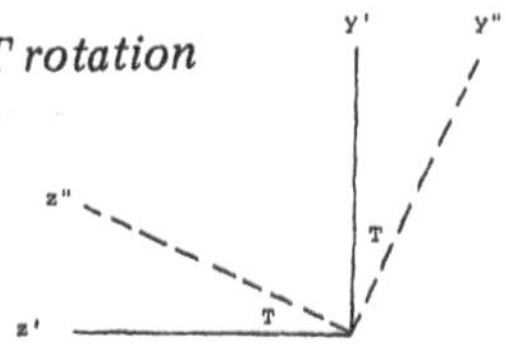

$$x' = x''$$ (C-15)

$$y' = y''\cos T + z''\sin T$$ (C-16)

$$z' = -y''\sin T + z''\cos T$$ (C-17)

or

$$x'' = x' = x\cos N_t - z\sin N_t$$ (C-18)

$$y'' = y'\cos T - z'\sin T = -x\sin N_t\sin T + y\cos T - z\cos N_t\sin T$$ (C-19)

$$z'' = y'\sin T + z'\cos T = x\sin N_t\cos T + y\sin T + z\cos N_t\cos T$$ (C-20)

$$x = x''\cos N_t - y''\sin N_t\sin T + z''\sin N_t\cos T$$ (C-21)

$$y = y''\cos T + z''\sin T$$ (C-22)

$$z = -x''\sin N_t - y''\cos N_t\sin T + z''\cos N_t\cos T$$ (C-23)

Now

$$x = R\cos E\sin A_z$$ (C-24)

$$y = R\sin E$$ (C-25)

$$z = R\cos E\cos A_z$$ (C-26)

or

$$R = [\,x^2 + y^2 + z^2\,]^{1/2}$$ (C-27)

$$\sin E = \frac{y}{R}$$ (C-28)

$$\tan A_z = \frac{x}{z}$$ (C-29)

and

$$x'' = R\sin\theta\cos\phi$$ (C-30)

$$y'' = R\sin\theta\sin\phi$$ (C-31)

$$z'' = R\cos\theta$$ (C-32)

or

$$R = [x''^2 + y''^2 + z''^2]^{1/2}$$ (C-33)

$$\cos\theta = \frac{z''}{R}$$ (C-34)

$$\tan\phi = \frac{y''}{x''} \tag{C-35}$$

From (C-1), (C-18), (C-24) and (C-26)

$$a = \frac{x''}{R} = \frac{x}{R}\cos N_t - \frac{z}{R}\sin N_t$$

$$a = \cos E \sin A_z \cos N_t - \cos E \cos A_z \sin N_t \tag{C-36}$$

and (C-2), (C-19), (C-24), (C-25) and (C-26)

$$\beta = \frac{y''}{R} = -\frac{x}{R}\sin N_t \sin T + \frac{y}{R}\cos T - \frac{z}{R}\cos N_t \sin T$$

$$\beta = -\cos E \sin A_z \sin N_t \sin T + \sin E \cos T - \cos E \cos A_z \cos N_t \sin T \tag{C-37}$$

From (C-25), (C-22), (C-1), (C-2) and (C-4)

$$\sin E = \frac{y}{R} = \frac{y''}{R}\cos T + \frac{z''}{R}\sin T$$

$$= \beta \cos T + \sqrt{1 - a^2 - \beta^2}\,\sin T$$

Or

$$E = \sin^{-1}\left[\beta \cos T + \sqrt{1 - a^2 - \beta^2}\,\sin T\right] \tag{C-38}$$

From (C-29), (C-21), (C-23), (C-1), (C-2) and (C-4)

$$\tan A_z = \frac{x}{z} = \frac{\frac{x''}{R}\cos N_t - \frac{y''}{R}\sin N_t \sin T + \frac{z''}{R}\sin N_t \cos T}{\frac{-x''}{R}\sin N_t - \frac{y''}{R}\cos N_t \sin T + \frac{z''}{R}\cos N_t \cos T}$$

$$A_z = \tan^{-1}\left[\frac{a\cos N_t - \beta \sin N_t \sin T + \sqrt{1-a^2-\beta^2}\,\sin N_t \cos T}{-a\sin N_t - \beta \cos N_t \sin T + \sqrt{1-a^2-\beta^2}\,\cos N_t \cos T}\right] \tag{C-39}$$

From (C-1), (C-2), (C-30) and (C-31)

$$a = \frac{x''}{R}\sin\theta\ \cos\phi \tag{C-40}$$

$$\beta = \frac{y''}{R} = \sin\theta \ \sin\phi \tag{C-41}$$

From (C-7) and (C-8)

$$\theta = \sin^{-1} \sqrt{a^2 + \beta^2} \tag{C-42}$$

$$\phi = \tan^{-1} \left[\frac{a}{\beta} \right] \tag{C-43}$$

From (C-38), (C-40) and (C-41)

$$E = \sin^{-1} \left[\sin\theta \sin\phi \cos T + \cos\theta \sin T \right] \tag{C-44}$$

From (C-39), (C-40) and (C-41)

$$A_z = \tan^{-1} \left[\frac{\sin\theta \cos\phi \cos N_t - \sin\theta \sin\phi \sin N_t \sin T + \cos\theta \sin N_t \cos T}{\sin\theta \cos\phi \sin N_t - \sin\theta \sin\phi \cos N_t \sin T + \cos\theta \cos N_t \cos T} \right] \tag{C-45}$$

From (C-34), (C-20), (C-24), (C-25) and (C-26)

$$\cos\theta = \frac{z''}{R} = \frac{x}{R} \sin N_t \cos T + \frac{y}{R} \sin T + \frac{z}{R} \cos N_t \cos T$$

$$\theta = \cos^{-1} \left[\cos E \sin A_z \sin N_t \cos T + \sin E \sin T + \cos E \cos A_z \cos N_t \cos T \right] \tag{C-46}$$

From (C-35), (C-18), (C-19), (C-24), (C-25) and (C-26)

$$\tan\phi = \frac{y''}{x''} = \frac{-\frac{x}{R} \sin N_t \sin T + \frac{y}{R} \cos T - \frac{z}{R} \cos N_t \sin T}{\frac{x}{R} \cos N_t - \frac{z}{R} \sin N_t}$$

$$\phi = \tan^{-1} \left[\frac{-\cos E \sin A_z \sin N_t \sin T + \sin E \cos T - \cos E \cos A_z \cos N_t \sin T}{\cos E \sin A_z \cos N_t - \cos E \cos A_z \sin N_t} \right] \tag{C-47}$$

List of Symbols

A = aperture size

A_z = azimuth angle

A' = $90° - A$

B_o = beamwidth at array normal — degrees

B_{θ_1} = beamwidth at scan angle θ_1

B' = $90° - B$

E = elevation

F = T = tilt angle

G = azimuth (arc) angle

H = maximum scan angle

K_o = 50.4 - 74 (depending on illumination taper)

K_x'' $= \dfrac{2\pi dx''}{\lambda}$

K_y'' $= \dfrac{2\pi dy''}{\lambda}$

L = length of linear array — meters

N_E = number of elements

N_t = angle between true north and plane $\perp$ to face of array

R $= \sqrt{x^2 + y^2 + z^2} = \sqrt{x''^2 + y''^2 + z''^2}$

T	= tilt angle of array from vertical
a	= element area
a_R	= element area for rectangular element spacing
a_Δ	= element area for equilateral element spacing
b	= beamwidth in sine space — radians
d	= interelement spacing
d_e	= effective element spacing in direction where grating lobes first appear
dx''	= interelement spacing in x'' direction
dy''	= interelement spacing in y'' direction
h	= side of circumscribed hexagon
ℓ	= element spacing for equilateral Δ configuration
$\ell_{x''}$	= element spacing along x'' coordinate
$\ell_{y''}$	= element spacing along y'' coordinate
m	= 1, 2, . . . M row
n	= 1, 2, . . . N column
x,y,z	= Cartesian coordinates in space
x'', y''	= Cartesian coordinates in plane of array
α	$= \sin A = \dfrac{x''}{R}$
β	$= \sin B = \dfrac{y''}{R}$
γ	$= \alpha - \beta$
δ	$= 2\beta$
θ, ϕ	= antenna spherical coordinates
θ_o	= half power beamwidth
θ_m	= maximum scan angle from array normal
λ	= transmission wavelength — meters
$\Delta\theta_{x''}$	$= \dfrac{2\pi dx''}{\lambda} \sin A$
$\Delta\theta_{y''}$	$= \dfrac{2\pi dy''}{\lambda} \sin B$
ϕ_{mn}	$= m\,\Delta\phi_{x''} + n\,\Delta\phi_{y''}$

ψ_γ	= differential phase shift between adjacent elements that forms grating lobe in γ direction
ψ_θ	= differential phase shift between adjacent elements to form beam in θ direction

ESR Tracking Considerations

Chapter Eight

8.1 ESR Tracking Characteristics — In ESRS s, the tracking loops are instrumented through proper manipulation of numbers representing observed errors, directed (predicted) beam positions and range gates, and the smoothed best estimates of target position, rate and acceleration.

For target amplitude, similar data quantities can be used to generate a smoothed measure of amplitude.

A set of recursive equations generate a sampled data servo loop within the computer. The equations are identical for all three coordinates and differ only in the values of the smoothing constants. Type II, III or Kalman filter designs are usually employed. The tracking algorithms operate to minimize the mean square error, weighting past data exponentially (in most designs).

The tracking and AGC operations of sampled data servos are equivalent to conventional tracking loops but are more flexible in that smoothing constant values can be much more readily altered, since they are represented by numbers in the tracking algorithms.

"Interpolative-null" tracking is characteristic of ESRS s. The electronically positioned beam is generally moved in discrete steps and each tracked target is illuminated at a low sampling rate. Therefore, the directed (predicted) beam position infrequently finds the target at the null of the beam and an interpolative correction increment relative to the null is required to provide accurate angular data. (This is in contrast to conventional tracking radars that null track at high sampling rates of 1 to 100 KHz.) The electronically positioned beam steps are in the order of 1/16 to 1/32 of a beamwidth and the target sampling rates range from 1/5 to 10 Hz. Even though only a limited amount of interpolation is needed, it still means that error slope calibration (with its accompanying angular error contribution) is required to provide a scale factor for the interpolative correction increment to be added to or subtracted from the center of the directed discrete beam position.

Low track sampling rates are characteristic of time-shared multifunction ESRS s. Time is conserved by employing the lowest sampling rates consistent with the tracking filter design and specified target dynamics. For example, the same time is required to track 100 targets at one look per second per target as is required to track 10 targets at 10 looks per second per target.

An ESR tracked target file is usually updated on a single-hit basis, which is consistent with low sampling rates. Monopulse, in addition to being the most accurate beam splitting technique, is also the only method that can be used for single hit operation. Single-hit angular error slope normalization, corresponding to taking the ratio Δ/Σ, may be obtained by one of two methods. The normalization may be performed in a limiting type or delay type three-channel monopulse receiver (the latter is the most accurate); or it may be performed by the system computer using the sum channel and two difference channel outputs, converted to digital format, from the monopulse receiver.

For high angular accuracy, it is recommended that the monopulse receiver include a built-in monitoring system employing a frequency coded RF pilot pulse to provide automatic and continuous phase and amplitude alignment of the three channels. The pilot pulse alignment can be accomplished during the dead time of the interpulse period.

The tracking loop is usually (but not always) designed in range and α, β sine space coordinates. Sine space coordinates simplify computations since (1) α and β are proportional to the array steering commands (i.e., the differential phase shift between the elements along the orthogonal axes of the array face) and (2) the monopulse error slope scale factor is invariant in sine space coordinates as a function of scan angle.

Track initiation after lock-on requires a separate filter design optimization of the tracking loop (except Kalman filters). The initial period of target tracking is based only on designated coordinates rather than on a history of past target motion. The designated coordinates appear to the filter as step inputs. Therefore, a wider bandwidth and a faster sampling rate are required than under steady-state operation to settle out transients more quickly and insure the maintenance ("tenacity") of track.

A target reacquisition mode is employed to retrieve lost targets. Loss of track is usually noted through an m out of n criterion performed continually on the tracked target return. A coast mode is recommended to maintain track during a temporary target fade or transmitter misfire. If m out of n consecutive returns (for example 3 out of 4) do not exceed the detection threshold, the target is assumed lost and the target reacquisition mode is initiated. Typically, the system performs a limited spiral search around the position where the target was last seen. If the target is reacquired, steady-state tracking (without reentering the track initiation mode) is continued. If the target is not reacquired the track file is emptied and made available for a new target track.

8.2 ESR Tracking Loop Filters — The selection of tracking loop filter designs for use in ESRS s is dependent on: target dynamics; accuracy (dynamic lag error); power and time availability (sampling rates for track maintenance) for both track initiation and steady-state tracking operation; and computation complexity (computer size and interpulse period time availability for computations). Type II, III and Kalman sampled data filter designs in their various forms are the most commonly employed in ESRS s. Evaluation of the above mentioned factors does not necessarily lead to a unique filter design selection. The recommended approach is to employ the simplest cost-effective filter design consistent with radar parameters that meets performance requirements.

Type II — velocity prediction (2 integrations) — The second order sampled data digital filter is the simplest and quite adequate design for many ESR tracking applications. In a second order system the sampled position data is processed to derive a measure of the rate of change in the position data. The smoothed position and rate data are the useful outputs of the network and form the basis for prediction of future position. The feedback compensation operates on angle (and range) observed errors to yield beam (and range gate) prediction commands (see Ref. 1). The networks are developed by minimizing the weighted squared error between the sampled data and a straight line approximating the data. An exponential weighting function is usually used although in some designs the last n samples are equally weighted. Accurate velocity data is required as quickly as possible after lock-on to implement a coast mode for protection against target loss due to fading or transmitter misfire.

The standard design approach is to start with a fixed digital filter configuration and optimize performance by adjusting the filter parameters (sampling rate, position and velocity smoothing constants) (see Ref. 2). The optimization can be performed either analytically or by simulation. The choice of smoothing constants is a compromise between filtering out high frequency noise and being able to follow evasive target maneuvers. Using as a performance index the sum of the squared errors from noise and target dynamics, and minimizing this index, it has been shown (see Ref. 3) that the optimum type II tracker has a slightly under-damped filter response. Further, for ramp inputs, the type II configuration is optimum among all linear time-invariant configurations. The limitation of this design approach is that target maneuvers which cannot be described by deterministic (i.e., ramp) inputs are not considered and the filter design can result only in a filter with fixed (time-invariant) parameters.

For lock-on, a higher sampling rate and a wider bandwidth (larger smoothing constant numbers) are employed than for steady-state operation. The sampling rate and bandwidth are decreased after a few seconds. There is apparently no analytical method to determine the best initial and final bandwidth and sampling rate values, in what increments to switch between them, when to switch, and what damping ratio constraint to impose. The filter parameters, in this time-invariant design, are chosen so that the position error is tolerable for worst case conditions.

Adaptive tracking loops can be implemented by changing the smoothing constants and sampling rates as function of the monitored observed error data, smoothed measured acceleration or range of the target in track.

In practice, simulation is employed extensively to optimize digital filter designs. Software simulators provide the means to optimize and evaluate digital filter designs for various types of target maneuvers, fluctuation models, noise and clutter environments. Hardware simulators expedite the debugging of transient and software interrupt timing problems. The finishing touches on a digital filter design are invariably performed in a field test environment with live targets.

For severe target dynamics, especially at close-in ranges and large scan angles, it is sometimes desirable to track in Cartesian coordinates instead of sine space coordinates to reduce the worse case target rate and acceleration components to acceptable design limits for a type II (or type III) system.

Type III — acceleration prediction (3 integrations) — The third order sampled data digital filter design is more complex and requires more computer computation time than a second order system. In a third order system the sampled position data is processed to measure position and velocity changes between samples. The smoothed position, velocity and acceleration data are the useful outputs of this network and are used to predict future position. The networks are developed by minimizing the weighted square error between the sampled data and a parabolic line approximating the data. The same types of weighting functions are used as in a second order system (see (Ref. 4).

In general (with some exceptions) a third order filter can be implemented at lower sampling rates than a second order filter and a third order filter can handle more severe target dynamics. In many applications a type II filter design is used for the angle coordinates and a type III filter design is used for the range coordinate.

All the previous remarks on a type II system concerning design optimization, adaptive tracking loops and simulation apply to a type III system as well.

Kalman Filters — Kalman tracking filters provide optimal feedback compensation and filtering in a digital computer (see Ref. 5). They are the most complex design of the types discussed but also provide the most advantages. A third order Kalman tracking filter design is similar to a type III in that smoothed position, velocity and acceleration output data is provided in all three coordinates based on least square fits. The important difference is that it provides the minimum rms error on these nine-state estimates given non-stationary radar measurement noise and no past track history. The gains (smoothing constants) are computed in real time as a function of measured signal-to-noise ratio and dynamic lag of the target in track. Automatic tracking loop parameter adjustments are made to cope with track initiation as well as steady-state track operation. Unlike the type II and type III configurations the Kalman design is not based on deterministic (i.e., ramp) inputs. The equations of motion of the object to be tracked serve as the basis for the Kalman filter predictor equations. The statistics of the noise and target fluctuation model determine the gains (smoothing constants) used in the Kalman filter corrector equations (see Refs. 6, 7, and 9).

The Kalman filter can cope with more severe target maneuvers at lower sampling rates than type II or type III filters. The use of Kalman filters is particularly advantageous where:

1. Very large number of targets (200-300) must be tracked

and power and time expenditure (sampling rate) must be kept very low (∿.2Hz) on each target.

2. Rapid estimation of target rates at lock-on is crucial because the observation time for a target trajectory is very short compared to impact point prediction time.
3. Target echo exhibits a wide variation in signal-to-noise ratio from pulse-to-pulse thus requiring an optimal time-varying filter configuration.
4. Target echo and radar error statistics are known a priori.

Even if the Kalman filter is not implemented, the design approach might be used to determine damping ratios and smoothing constants that can then be applied to the design of type II or type III filters with piece-wise constant filter parameters.

In many applications, the computational complexity of the Kalman filter can be significantly reduced by propagating co-variance matrices in RPCC (radar principal Cartesian coordinates) (see Ref. 8). In this coordinate system the nine-state filter is essentially decoupled into 3 three-state filters. Thus the covariance matrix can be propagated as three separate 3 x 3 matrices instead of one 9 x 9 matrix, so that the nine-state decoupled filter is the computational equivalent of three three-state filters. The computational complexity is 5 to 10 times less than a fully coupled nine-state filter.

8.3 Design Equations for Type II and Type III Digital Filters —

The recursive tracking equations and design formulas for second and third order systems are tabulated for a critically damped system. The type II and type III tracking loops operate to minimize the mean square error for a straight line and parabolic fit respectively, weighting past data exponentially (Refs. 1 and 4). An example is given in the next section to provide familiarity with their usage.

Design Formulas

Type II —

$$\epsilon_a = \frac{at^2}{\beta_{CD}} \tag{8.1}$$

$$\alpha_{CD} = 1 - \theta^2 \tag{8.2}$$

Type III —

$$\epsilon_j = \frac{jt^3}{\gamma_{CD}} \tag{8.10}$$

$$\alpha_{CD} = 1 - \theta^3 \tag{8.11}$$

$$\beta_{CD} = (1-\theta)^2 \quad (8.3)$$

$$(2\alpha_{CD} + \beta_{CD}) \leqslant 4 \quad (8.4)$$

$$0 < \alpha_{CD} < 1 \quad (8.5)$$

$$0 < \beta_{CD} < 4 \quad (8.6)$$

$$\epsilon_{\substack{a \\ j}} \leqslant \begin{Bmatrix} \dfrac{\tau}{2} \\ \\ \dfrac{\theta_o}{2} \end{Bmatrix} \quad (8.7)$$

$$f(\theta) = \theta^n \quad (8.8)$$

$$-1 < \theta < 1 \quad (8.9)$$

$$\beta_{CD} = \frac{3}{2}(1+\theta)(1-\theta)^2 \quad (8.12)$$

$$\gamma_{CD} = (1-\theta)^3 \quad (8.13)$$

$$0 < \alpha_{CD} < 2 \quad (8.14)$$

$$0 < \beta_{CD} < \frac{16}{9} \quad (8.15)$$

$$0 < \gamma_{CD} < 8 \quad (8.16)$$

where:

$f(\theta) = \theta^n$ = exponential weighting function

n = n^{th} sample, n = 1, 2, . . .

α_{CD} = position smoothing constant for a critically damped system

β_{CD} = velocity smoothing constant for a critically damped system

γ_{CD} = acceleration smoothing constant for a cirtically damped system

ϵ_a = acceleration lag between samples

ϵ_j = jerk lag between samples

a = acceleration/deceleration of target

j = jerk of target

t = time interval between samples
τ = equivalent pulse width
θ = weighting factor
θ_o = 3 dB beamwidth

Recursive Tracking Equations

Type II —

$$\epsilon_n = X_n - X_{P_n} \quad (8.17)$$

$$\overline{X}_n = X_{P_n} + \alpha_{CD}\epsilon_n \quad (8.18)$$

$$\overline{\dot{X}}_n = \overline{\dot{X}}_{n-1} + \frac{\beta_{CD}}{t}\epsilon_n \quad (8.19)$$

$$X_{P_{n+1}} = \overline{X}_n + t\overline{\dot{X}}_n \quad (8.20)$$

Type III —

$$\epsilon_n = X_n - X_{P_n} \quad (8.21)$$

$$\overline{X}_n = X_{P_n} + \alpha_{CD}\,\epsilon_n \quad (8.22)$$

$$\overline{\dot{X}}_n = \overline{\dot{X}}_{n-1} + t\,\overline{\ddot{X}}_{n-1} + \frac{\beta_{CD}}{t}\epsilon_n \quad (8.23)$$

$$\overline{\ddot{X}}_n = \overline{\ddot{X}}_{n-1} + \frac{\gamma_{CD}}{t^2}\epsilon_n \quad (8.24)$$

$$X_{P_{n+1}} = \overline{X}_n + t\,\overline{\dot{X}}_n + \tfrac{1}{2}t^2\,\overline{\ddot{X}}_n \quad (8.25)$$

where:

ϵ_n = position error on n^{th} sample
X_n = observed target position on n^{th} sample
X_{P_n} = predicted target position on n^{th} sample
$\overline{X}_n$ = smoothed target position on n^{th} sample
α_{CD} = position smoothing constant for a critically damped system
β_{CD} = velocity smoothing constant for a critically damped system
γ_{CD} = acceleration smoothing constant for a critically damped system
t = time interval between samples

$\bar{\dot{X}}_n$ = smoothed target velocity on n^{th} sample

$\bar{\ddot{X}}_n$ = smoothed target acceleration on n^{th} sample

8.4 Use of Tracking Loop Design Equations —

Illustrative Example — A radar with a 1° beamwidth and a 10 MHz bandwidth waveform is required to track a missile target from 100 nm maximum range to 10 nm minimum range. The missile dynamic characteristics are:

v_{max} = Mach 2

a_{max} = 15 g

j_{max} = 2 g/s

a. Determine the longest sampling interval than can be used for track maintenance for both type II and type III filter designs.

b. Determine the filter designs, sampling rates, smoothing constants and weighting function (as a first approximation) for range and angle tracking.

Part (a) —

Range — Type II

$$t = \left[\frac{\beta_{CD}\,\epsilon_a}{a}\right]^{1/2} \qquad \text{[Eq. (8.1)]}$$

$$\beta_{CD} = 4\ (\text{max}) \qquad \text{[Eq. (8.6)]}$$

$$\epsilon_a = \frac{\tau}{2} = \frac{1}{2B} = \frac{1}{2 \times 10^7} = .05\mu s \times 150\ m/\mu s = 7.5m$$

[Eq. (8.7)]

$$a = \frac{15 \times 32.2}{3.281} = 147.21\ m/s^2$$

$$t = \left[\frac{4 \times 7.5}{147.21}\right]^{1/2} = .45\ s$$

ANSWER

Range — Type III

$$t = \left[\frac{\gamma_{CD}\,\epsilon_j}{j}\right]^{1/3} \qquad \text{[Eq. (8.10)]}$$

$$\gamma_{CD} = 8\ (\text{max}) \qquad \text{[Eq. (8.16)]}$$

$$\epsilon_j = 7.5\ \text{m}$$

$$j = \frac{2 \times 32.2}{3.281} = 19.63\ \text{m/s}^3$$

$$t = \left[\frac{8 \times 7.5}{19.63}\right]^{1/3} = \mathbf{1.45}\ \text{s}$$ ANSWER

Angle — Type II

$$t = \left[\frac{\beta_{CD}\,\epsilon_a}{\ddot{\phi}}\right]^{1/2} \qquad \text{[Eq. (8.1)]}$$ where $\ddot{\phi}$ = angular acceleration.

$$\beta_{CD} = 4\ (\text{max}) \qquad \text{[Eq. (8.6)]}$$

$$\epsilon_a = \frac{\theta_o}{2} = \frac{.01745}{2} = .008727 \qquad \text{[Eq. (8.7)]}$$

$$\ddot{\phi}_1 = \frac{a}{R_{min}} = \frac{147.21}{10 \times 1852} = .007948$$

$$\ddot{\phi}_2 = .65\left(\frac{v}{R_{min}}\right)^2 = .65 \times \left(\frac{2 \times 340}{18520}\right)^2 = .000876$$

Use $\ddot{\phi} = \ddot{\phi}_1$ since $\ddot{\phi}_1 > \ddot{\phi}_2$

$$\therefore\ t = \left[\frac{4 \times .008727}{.007948}\right]^{1/2} = 2.1\ \text{s}$$ ANSWER

Angle — Type III

$$t = \left[\frac{\gamma_{CD}\,\epsilon_j}{\dddot{\phi}}\right] \qquad \text{[Eq. (8.10)] where } \dddot{\phi} = \text{angular jerk}$$

$$\gamma_{CD} = 8 \text{ (max)} \qquad \text{[Eq. (8.16)]}$$

$$\epsilon_j = .008727$$

$$\dddot{\phi} = \frac{j}{R_{min}} = \frac{19.63}{18520} = .001060$$

$$\therefore t = \left[\frac{8 \times .008727}{.001060}\right]^{1/3} = 4.04 \text{ s} \qquad \text{ANSWER}$$

Part (b) — The results of part (a) indicate that to a first approximation it is feasible to sample at a one second interval if a type III system is employed for range and a type II system is adequate for angle.

$\therefore$ *Range — Type III*

$$\gamma_{CD} = \frac{jt^3}{\epsilon_j} \qquad \text{[Eq. (8.10)]}$$

$$j = 19.63 \text{ m/s}^3$$

$$\epsilon_j = 7.5 \text{ m}$$

$$t = 1\text{s}$$

$$\gamma_{CD} = \frac{19.63 \times 1^3}{7.5} = 2.617$$

$$\theta = 1 - \gamma^{1/3} = 1 - 2.617^{1/3} = -.3781 \quad \text{[Eq. (8.13)]}$$

$$\therefore\ f(\theta) = (-.3781)^n \qquad \text{[Eq. (8.8)]}$$

$$\alpha_{CD} = 1 - \theta^3 = 1 + .3781^3 = 1.054 \qquad \text{[Eq. (8.11)]}$$

$$\beta_{CD} = \frac{3}{2}(1+\theta)(1-\theta)^2 = \frac{3}{2} \times .6219 \times 1.3781^2 = 1.77$$

Angle — Type II

$$\beta_{CD} = \frac{\ddot{\phi}\, t^2}{\epsilon_a} \qquad \text{[Eq. (8.1)]}$$

$$t = 1s$$

$$\ddot{\phi} = .007948 \text{ rad/s}^2$$

$$\epsilon_a = .008727 \text{ rad}$$

$$\beta_{CD} = \frac{.007948 \times 1^2}{.008727} = .911$$

$$\theta = 1 - \sqrt{\beta} = 1 - \sqrt{.911} = .04568 \qquad \text{[Eq. (8.3)]}$$

$$f(\theta) = .0457^n \qquad \text{[Eq. (8.8)]}$$

$$\alpha_{CD} = 1 - \theta^2 = 1 - .04568^2 = .9979$$

Summary

	Type	t	$f(\theta)$	α_{CD}	β_{CD}	γ_{CD}
Range	III	1.0s	$-.3781^n$	1.0540	1.770	2.617
Angle	II	1.0s	$.0457^n$	0.9979	0.911	---------

References

[1] Dale M. Jahn, "Smoothing of Sampled Data," *Sperry-Rand Internal Report #5284.7345*, February 4, 1958.

[2] F.E. Daum, "The Application of Kalman Filtering Techniques to Radar Signal Processing," *Raytheon Technical Memo #EM70-0584*, October 1, 1970.

[3] T.R. Benedict and G.W. Bordner, "Synthesis of an Optimum Set of Radar Track-While-San Smoothing Equations," *IRE Trans. on Automatic Control*, July 1962.

[4] A.H. Miller, "Smoothing Constants for a Type III System," *Sperry-Rand Technical Memo*, October 31, 1967.

[5] R.E. Kalman, "A New Approach to Linear Filtering and Prediction Problems," *Journal of Basic Engineering*, March 1960.

[6] H.W. Sorenson, "Kalman Filtering Techniques in Advances in Control Systems," by C.T. Leondes, Vol. 3, Academic Press, 1966.

[7] A.E. Bryson and Y.C. Ho, "Optimal Filtering and Prediction," Chapter 12 of *Applied Optimal Control*, Blaisdell Publishing Company, 1969.

[8] G.W. Johnson and D.H. Winfield and K.R. Brown, "An Improved Algorithm for Safeguard Missile Tracking and Malfunction Detection," *IBM report to BTL under Contract #606055*, October 30, 1970.

[9] A.S. Locke, "Guidance," D. Van Nostrand Co., Princeton, N.J., 1955.

List of Symbols

B = signal bandwidth

X_n = observed target position on n^{th} sample

X_{P_n} = predicted target position on n^{th} sample

$\overline{X}_n$ = smoothed target position on n^{th} sample

$\overline{\dot{X}}_n$ = smoothed target velocity on n^{th} sample

$\overline{\ddot{X}}_n$ = smoothed target acceleration on n^{th} sample

a = acceleration/deceleration of target

a_{max} = maximum acceleration

$f(\theta)$ = θ^n = exponential weighting function

j = jerk of target

j_{max} = maximum jerk

n = nth sample $n = 1, 2, \ldots$

t = time interval between samples

v_{max} = maximum velocity

α_{CD} = position smoothing constant for a critically damped system

β_{CD} = velocity smoothing constant for a critically damped system

γ_{CD} = acceleration smoothing constant for a critically damped system

ϵ_a = acceleration lag between samples

ϵ_j	= jerk lag between samples
ϵ_n	= position error on n^{th} sample
θ	= weighting factor
θ_o	= 3 dB beamwidth
τ	= equivalent pulse width
Δ	= difference signal — monopulse
Σ	= sum signal — monopulse
$\ddot{\phi}$	= angular acceleration
$\ddot{\phi}_1$	= radial angular acceleration
$\ddot{\phi}_2$	= tangential angular acceleration
$\dddot{\phi}$	= angular jerk

Accuracy and Resolution

Chapter Nine

Radar measurement accuracy is based on the estimation of parameters of statistical distributions (probability measures) in an environment of noise, clutter and nonhomogeneous propagation media. The differences among accuracy, precision and resolution are briefly discussed to eliminate any confusion in the understanding and application of these concepts. The significance of radar parameters with respect to measurements and derivation of accuracy formulas has been well covered in the literature (see references) for both conventional and ES radar systems.

The intent here is to focus on the error sources that determine the accuracy of ESRSs. The basic error contributors and estimation formulas are conveniently summarized in tabular form. Differences exist in measurement-accuracy formulas due to:

1. Types of noise — white or colored; additive or multiplicative
2. Small or large signal-to-noise ratios
3. Types of waveforms
4. Beam splitting techniques and array illumination tapers
5. Types of signal processing

The expressions included in the tabulation apply to most ESR designs. Other formulations for special cases are treated in the referenced literature. The error sources unique to ESRS s are readily recognizable in the tabulation. Typical magnitudes of the errors are provided to guide the ESRS designer in the allocation of error budgets.

9.1 Definitions —

Accuracy — Accuracy is defined as freedom from mistake or error. Radar measurement accuracy of target position is based on the total deviation between the actual target location in space (3 coordinates) and that location as reported by the radar processed output.

Accuracy includes:

1. Survey bias errors — geodetic uncertainties of longitude and latitude
   — local surveys
2. Atmospheric propagation errors — refractive
   — non homogeneous media

Accuracy is expressed as the square root of the mean-square error, i.e., rms error. All the accuracy formulas represent the accuracy of parameter measurement when there is no other object near the target to affect the measurement (a background of distributed clutter sources may exist).

Precision — Precision relates to exactness. It represents the repeatability and measureable radar error variance and unknown bias errors. Radar precision is primarily a measure of circuit stability, thermal noise and instrumentation errors that have not or cannot be compensated by calibration or removed by bias corrections. In radar usage precision is part of accuracy and is an indicator of the inherent capability of the radar.

Resolution — Resolution refers to the ability of the radar to separate two (or more) targets that are close together, and to determine in range, angle or doppler frequency (radial velocity) that there are two instead of one. Resolution specifications are given in terms of minimum distance between targets and maximum dynamic range of target size.

9.2 ESR Sources of Error — The estimation formulas and/or contribution for the basic error sources in ESRs are summarized in Tables 9.1 to 9.3. The reference sources are indicated in parentheses. In some cases the formulas have been modified slightly to generalize their usage. In one or two cases (i.e., dynamic lag and monopulse signal processing) specific designs were assumed both to indicate the error contributing parameters and as a reminder that these error sources exist and are to be included in an ESRS error budget. Typical magnitudes of the error contributions are indicated in the last column of each table. The thermal noise errors in the measurement of amplitude and phase are included at the end of this section without tabulation since the other amplitude and phase sources can be readily identified from the other tabulations.

Table 9.1
ESR Sources of Error-Angle

	Error Source	Equation and/or Contribution	Typical Magnitudes (RMS)
1.	Antenna Pointing (14,15)	Manufacturing tolerances (mechanical & electrical), temperature, wind, solar, gravity, mount deflection (if mount is used) and array face deflection. $\sigma_b = \frac{c\theta_o}{2\sqrt{N_e}}$ (9.1) σ_b = beam pointing error — millisines. ϵ = rss of phase errors at each antenna element — radians (due to: element spacing, axial displacement & mismatch; phase shifter tolerances & mismatch; cables; corporate feed). θ_o = 3 dB beamwidth — millisines N_e = total number of active elements.	.001 to .003 Beamwidth
2.	Phase Shifter Quantization (16)	$\sigma_{\phi Q} = \frac{2.6\,\theta_o}{N\,2^P}$ (9.2) $\sigma_{\phi Q}$ = phase quantization beam pointing error — millisines. P = number of phase shifter bits. N_L = number of elements in linear array.	.001 to .005 Beamwidth
3.	Data Quantization	$\sigma_{Q_A} = \frac{2\pi}{\sqrt{12}\,2^Q}$ (9.3) or $\sigma_{Q_A} = \frac{LSB}{\sqrt{12}}$ (9.4) σ_{Q_A} = angle quantization error — radians. Q = number of computer bits (usually $\geq$ 16). LSB = least significant bit value — radians.	.001 to .03 Mils.
4.	A/D Conversion	Errors of analog sampling in digital conversion due to circuitry and contact potential.	.01 to .05 Mils.
5.	Thermal Noise (4,5)	$\sigma_{T_A} = \frac{\theta_o}{K_M\sqrt{2n\,S/N}}$ (9.5) σ_T = angle thermal error — millisines. n = number of hits smoothed. S/N = single-hit signal-to-noise ratio. K_M = normalized monopulse slope (5) = 1.89 typical value. = 1.2 to 2.0 normal range.	1/5 to 1/200 Beamwidth
6.	Instrumentation (17)	Error due to monopulse feedhorn and receiver amplitude and phase errors. Assume 4 horn monopulse with single-hit normalization receiver. $\sigma_I = \left[1 - G_1 \frac{G_2\cos(\alpha_1 + \alpha_2 + \alpha_3) + \sin(\alpha_1 + \alpha_3)}{G_2\cos(\alpha_2 + \alpha_3) - \sin\alpha_3}\right] S_o$ (9.6) σ_I = angle instrumentation error — millisines. $S_o = \frac{\theta_o}{K_M\sqrt{G}}$ = reciprocal of difference slope — mr/gain differential. $\sqrt{G}$ = antenna voltage gain. G_1 = precomparator amplitude error — gain ratio. α_1 = precomparator phase error — degrees. G_2 = post comparator amplitude error — gain ratio. α_2 = post comparator phase error — degrees. α_3 = post post comparator (limiter) phase error — degrees.	1/100 to 1/400 Beamwidth
7.	Dynamic Lag (18)	Assume type II sampled data tracking loop, critically damped minimizes mean square error and weights past data exponentially. $\sigma_{DL_A} = \frac{\ddot{\phi} T_s^2}{\beta_{CD}}(1 - \alpha_{CD})$ (9.7) σ_{DL_A} = angle dynamic lag error — radians. $\ddot{\phi}$ = maximum angular acceleration — radians/s^2. T_s = sample period — s. α_{CD} = position smoothing constant. β_{CD} = velocity smoothing constant.	0.1 to 0.5 Mils.

8.	Scan	(19)	Error due to monopulse slope variation with scan angle from array normal — on a least mean square fit basis — produces a zero crossing (null bias) error and a total error within the 3 dB beamwidth.	Null 1/50 to 1/100 **Beamwidth** Total **1/25 to 1/50** Beamwidth
9.	Multipath	(4,5)	$\sigma_{M_A} = \dfrac{\theta_o \rho}{\sqrt{8\, G_{SL}}}$ (9.8) σ_{M_A} = angle multipath error — millisines ρ = surface reflectivity. G_{SL} = mainlobe to sidelobe ratio in direction of reflected ray, or = rms sidelobe level.	1/200 to 1/1000 Beamwidth
10.	Glint	(4,5)	$\sigma_{G_A} = \dfrac{.35\, L_X}{R}$ (9.9) σ_{G_A} = angle glint error — radians. L_X = cross range target span — m (cross range distance between major scatterers). R = range to target — m.	.02 to .5 Mils.
11.	Propagation	(4)	All errors are given for average tropospheric conditions and may vary by a factor of 3 or more. Portions of errors may be eliminated by calibration in the region near the radar site against objects whose positions are accurately known. Elevation error on path through entire troposphere: $\Delta E = N_s \cot E \quad (E > 5^\circ)$ (9.10) $\Delta E_{max} \approx 15\ mr_1 \quad (E \sim 0^\circ)$ (9.11) Angular error for "frozen atmosphere": $\sigma_\theta = \dfrac{L}{V_a} \left[\int_0^\infty H'^2(f)\, \overline{W}(f)\, df \right]^{1/2}$ μrad. (9.12) Angle error for aperture width W: $\sigma_\theta \approx .44 \dfrac{L}{\sqrt{W}}$ (μrad) (9.13) Azimuth error for baseline b: $\sigma_\theta \approx .4 \dfrac{L}{\sqrt{b}}$ (μrad.) $(b \ll L)$ (9.14) Elevation error for baseline system $R \approx L$: $\sigma_{E_o}^2 = \sigma_\theta^2 + (16 \cot E)^2$ (μrad.) (9.15) Elevation error for long range $R \gg L$: $\sigma_{E_o}^2 = \sigma_\theta^2 + (2 \cot E)^2$ (μrad.) (9.16) ΔE = elevation bias error — mr. N_s = surface refractivity. E = ray elevation angle at surface — degree. σ_θ = angular propagation error under average conditions — μrad. L = path length within troposphere — m. V_a = average wind speed — m/s. $H'^2(f)$ = differentiator function. $\overline{W}(f)$ = refractivity spectrum for average conditions. W = aperture width — m. b = baseline length — m. σ_E = elevation error from baseline system — μrad.	0 - 300μrad.

Table 9.2
ESR Sources of Error-Range

	Error Source	Equation and/or Contribution	Typical Magnitudes (RMS)
1.	Data Quantization	$\sigma_{Q_R} = \frac{R_M}{\sqrt{12}\, 2^Q}$ (9.17) or $\sigma_{Q_R} = \frac{LSB}{\sqrt{12}}$ (9.18) σ_{Q_R} = range quantization error — m. R_M = maximum instrumented range — m. Q = number of computer bits (usually $\geqslant 20$). LSB = least significant bit value — m.	.01 to 0.1 m
2.	Pulse Jitter	Transmitter time synchronization pulse jitter. $\sigma_{P_R} = \frac{c\,\Delta t_s}{2\sqrt{12}}$ (9.19) σ_{P_R} = range pulse jitter error — m. Δt_s = maximum pulse jitter — s. c = 3×10^8 m/s — velocity of light.	2 to 10 ns or .09 to .43 m
3.	Range Clock	Range clock frequency quantization: (clock instability is negligible ~ 1 in 10^7 for crystal) $\sigma_c = \frac{c}{2\sqrt{12}\, f_c}$ (9.20) σ_c = range clock error — m. f_c = clock frequency — Hz.	1.25 to 12.5 ns or .05 to .5 m
4.	FM Waveform (6)	For a linear FM signal waveform with a positive slope, a doppler shift causes a relative time shift at the matched filter that produces a range error. This error can be compensated for in the tracking computer by observing the target position over a space of N pulses and obtaining the range rate (assuming no acceleration). $\sigma_{FM} = \frac{T_o V_M f_o}{2\sqrt{12}\,\Delta f}$ (9.21) σ_{FM} = range FM error — m (for linear FM with rectangular envelope waveform). T_o = signal time duration — s. V_M = maximum target velocity — m/s. f_o = transmission frequency — Hz. Δf = signal bandwidth — Hz. (FM deviation during time T_o).	.15 to 15 m
5.	A/D Conversion	Errors of analog sampling in digital conversion due to circuitry and contact potential.	.1 to .5 m
6.	Thermal	$\sigma_{T_R} = \frac{C\,\tau_e}{2\sqrt{2nS/N}}$ (9.21) σ_{T_R} = range thermal error — m. τ_e = equivalent pulse length — s. $(\tau_e = \frac{1}{B})$ n = number of hits smoothed. S/N = single-hit signal-to-noise ratio.	1/5 to 1/200 of range resolution.
7.	Dynamic Lag (18)	Assume type II sampled data tracking loop, critically damped, minimizes mean square error and weights past data exponentially. $\sigma_{DL_R} = \frac{a T_s^2}{\beta_{CD}} (1 - \alpha)$ (9.23) σ_{DL_R} = range dynamic lag error — m. a = maximum acceleration m/s^2. T_s = sampled period — s. α_{CD} = position smoothing constant. β_{CD} = velocity smoothing constant.	0.1 to 10 m
8.	Multipath (4,5)	$\sigma_{M_R} = \frac{c\,\rho\,\tau_e}{2\sqrt{8}\, G_{SL}}$ (9.24) σ_{M_R} = range multipath error — m. ρ = surface reflectivity. G_{SL} = mainlobe to sidelobe ratio in direction of reflected ray. or = rms sidelobe level.	1/200 to 1/1000 of range resolution.
9.	Glint (4,5)	$\sigma_{G_R} = .35\, L_R$ (9.25) σ_{G_R} = range glint error — m. L_R = radial target span — m. (radial distance between major scatterers).	0.1 to 10 m

10.	Propagation	All errors are given for average tropospheric conditions and may vary by a factor of 3 or more. Portions of errors may be eliminated by calibration in the region near the radar site (4) against objects whose positions are accurately known. Range error on path through entire troposphere: $\Delta R_b = .007\ N_s \csc E\ (m)\ (E > 5°)$ (9.26) $\Delta R_{max} \approx 100$ m (9.27) Range fluctuation for average conditions: $\sigma_r \approx 1.5 \times 10^{-5}\ L$ (9.28) Apparent range error in restricted observation period, t_o: $\sigma_{ro} \approx 1.4 \times 10^{-7}\ L^{0.5}\ t_o^{\ 0.75}\ (t_o < 10^4\ s)$ (9.29) Range error after refractive corrections: $\sigma_{ro} \approx 0.1\ \sigma_r$ (9.30) ΔR_b = range bias error — m. N_s = surface refractivity. E = ray elevation angle at surface — degrees. σ_r = range fluctuation for average conditions — m. L = pathlength within troposphere — m. σ_{ro} = apparent range fluctuation observed in restricted interval — m. t_o = observation time — s. σ_{rc} = range fluctuation after refractivity — m.	0 to 5 m

Table 9.3

Sources of Error-Velocity

	Error Source	Equation and/or Contribution	Typical Magnitudes (RMS)
1.	Quantization	$\sigma_{QV} = \dfrac{V_m}{\sqrt{12}\ 2^Q}$ (9.31) σ_Q = velocity quantization error — m/s V_m = maximum instrumented velocity — m/s Q = number of computer bits (usually ⩾16).	.01 to .1 m/s
2.	Pulse Jitter	Transmitter time synchronization pulse jitter: $\sigma_{PV} = \dfrac{c\,\Delta t_s}{2\sqrt{6}\ T_o}$ (9.32) σ_{PV} = velocity pulse jitter error — m/s Δt_s = maximum pulse jitter between samples — s. c = 3×10^8 m/s — velocity of light. T_o = time between looks — s or look time on target.	.06 to .3 m/s
3.	Doppler Discriminator Stability & A/D Conversion	Errors due to circuitry, contact potential and oscillator frequency stability.	.15 to .7 m/s
4.	Thermal	Doppler (10) $\sigma_{TV} = \dfrac{\lambda}{2T\sqrt{2S/N}}$ (9.33) σ_{TV} = velocity thermal error — m/s λ = transmission wavelength — m T_d = signal time duration — s S/N = single hit signal-to-noise ratio.	0.5 to 50 m/s
		Non-coherent (9.34) $\sigma_{TV} = \dfrac{c\tau_e}{2\ n^{3/2} T_o \sqrt{S/N}}$ σ_{TV} = velocity thermal error — m/s τ_e = equivalent pulse length — s n = number of hits smoothed	1.5 to 15 m/s

5.	Dynamic Lag	Assume type II sample data tracking loop, critically damped, minimizes mean square error and weights past data exponentially. $$\sigma_{DL_V} = \frac{\sqrt{2}\, a T_s}{\beta_{CD}} (1 - \alpha) \quad (9.35)$$	σ_{DL_V} = velocity dunamic lag error — m/s. a = maximum acceleration — m/s^2. T_s = sample period — s. α_{CD} = position smoothing constant. β_{CD} = velocity smoothing constant.	0.15 to 15 m/s
6.	Multipath	(4) $$\sigma_{M_V} = \frac{h \rho \dot{E}}{\sqrt{2 G_{SL}}} \quad (9.36)$$	σ_{M_V} = velocity multipath error — m/s h = radar antenna height — m ρ = surface reflectivity $\dot{E}$ = elevation angle rate of target — radians/s. G_{SL} = rms sidelobe ratio	.005 to .06 m/s
7.	Glint	(4) $$\sigma_{G_V} = .35\, L_X \omega \quad (9.37)$$	σ_{G_V} = velocity glint error — m/s L_X = cross range target span — m (cross range distance between major scatterers). ω = target rotation range — radians/s	0.5 to 15 m/s
8.	Propagation	Doppler velocity error due to change in range delay in troposphere that is proportional to local refractivity index. $$\sigma_N = V_T N_t \times 10^{-6} \quad (9.38)$$	σ_N = velocity propagation error — m/s V_T = target radial velocity N_t = tropospheric refractivity in vicinity of target	0 - .01 m/s
		Doppler velocity error due to ray bending in troposphere: $$\sigma_{VE} = \dot{E}\, L\, \sigma_{Az} \quad (9.39)$$ $$\sigma_{VAz} = \dot{A}z\, L\, \sigma_{Az} \quad (9.40)$$	σ_{VE} = elevation coordinate velocity propagation error — m/s σ_{VA_z} = azimuth coordinate velocity propagation error — m/s $\dot{E}$ = elevation angle rate of target — radians/s. $\dot{A}_z$ = azimuth angle rate of target — radians/s. L = length of tropospheric portion of path — m (below 5 Km altitude). σ_E = unknown or uncorrected portion of tropospheric bias error ΔE — radians σ_{Az} = unknown or uncorrected portion of tropospheric bias error in azimuth — radians.	
		Range rate error due to tropospheric fluctuations: $$\sigma_{TF} \approx .33 \times 10^{-12}\, v_d\, 3/2\, \bar{f}_d\, ^{1/2} L \quad (9.41)$$	σ_{TF} = range rate propagation error — m/s. v_d = drift velocity of air mass — m/s $\bar{f}_d$ = smoothed doppler frequercy ∿ 1 Hz.	

Thermal noise error in measurement of amplitude [Refs. 2 and 3]

$$\sigma_A = \frac{A}{\sqrt{2\,S/N}} \qquad (9.42)$$

Phase

Thermal noise error in measurement of phase [Refs. 2 and 3]

$$\sigma_\phi = \frac{1}{\sqrt{2\,S/N}} \qquad (9.43)$$

where:

σ_A = amplitude error — volts

A = amplitude — volts

σ_ϕ = phase error — radians

9.3 Resolution — Resolution is the ability of the radar to separate two (or more) targets that are close together in at least one of the coordinates the radar is measuring (unless customer specifications contain resolution requirements in each coordinate).

The design criterion for maximum sidelobe level of the antenna pattern and/or maximum residue level of the signal waveform is based on the same rationale as employed to determine the MTI improvement factor requirement (see Fig. 6.1 and section 6.2).

Angle

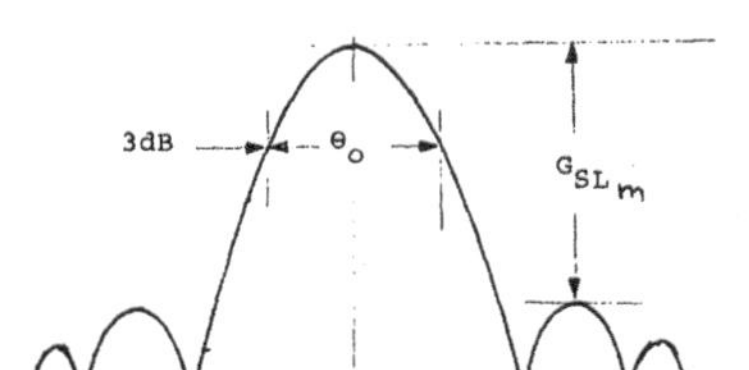

$$\Delta\theta = K_R\,\theta_o \qquad (9.44)$$

$$G_{SL_{m_{dB}}} = \tfrac{1}{2}\left[(S/N)_{o_{dB}} + \sigma_{\Delta_{dB}} + 6_{dB}\right] \qquad (9.45)$$

where:

$\Delta\theta$	= angle resolution — millisenes
K_R	= 1 to 2 (1.5 typical) resolution factors
θ_o	= half power beamwidth — millisines
$(S/N)_o$	= single pulse detectability factor
G_{SL_m}	= maximum sidelobe level of antenna pattern
σ_Δ	= maximum dynamic range of target size

Range

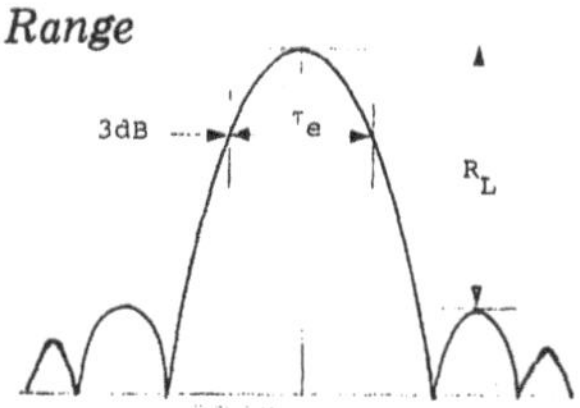

$$\Delta R = \frac{K_R c\tau_e}{2} \tag{9.46}$$

$$R_L = \left[(S/N)_{o_{dB}} + \sigma_{\Delta_{dB}} + 6_{dB}\right] \tag{9.47}$$

where:

ΔR	= range resolution — m
τ	= equivalent half power pulse width — s
c	= 3 x 10^8 m/s velocity of light
R_L	= maximum residue level of signal waveform

Doppler Velocity

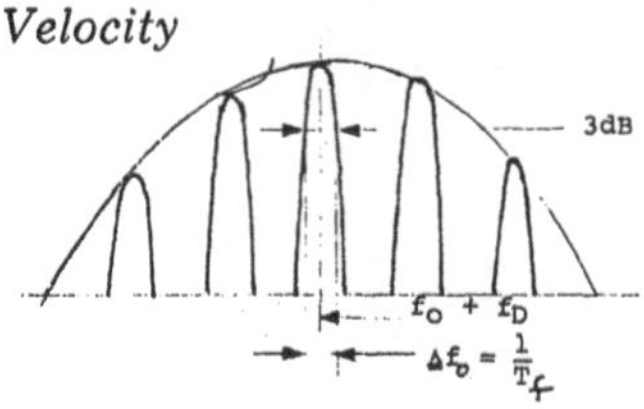

$$\Delta v = \frac{K_R c}{2T_f f_o} \tag{9.48}$$

where:

Δv	= doppler velocity resolution — m/s
Δf_o	= half power spectral line width — Hz
f_o	= transmission frequency — Hz
T_f	= finite time on target — s or finite length of pulse or finite length of pulse train
f_d	= doppler frequency

References

[1] P.M. Woodward, "Probability and Information Theory, with Applications to Radar," Pergamon Press, 1953.

[2] M.I. Skolnik, "Radar Handbook," McGraw-Hill, 1970.

[3] M.I. Skolnik, "Introduction to Radar Systems," McGraw-Hill, 1962.

[4] D.K. Barton and H.R. Ward, "Handbook of Radar Measurement," Prentice-Hall, 1969.

[5] D.K. Barton, "Radar Systems Analysis," Prentice-Hall, 1964.

[6] C.E. Cook and M. Bernfield, "Radar Signals," Academic Press, 1967.

[7] P. Swerling, "Maximum Angular Accuracy of a Pulsed Search Radar," *Proc. IRE*, pp 1146-1155, Sept. 1956.

[8] P. Swerling, "Detection of Radar Echoes in Noise Revisited," *IEEE Trans.* Vol. IT-12, pp 348-361, July 1966.

[9] L.E. Brennan, "Angular Accuracy of a Phased Array Radar," *IRE Trans.*, Vol. AP-9, pp 268-275, May 1961.

[10] R. Manasse, "Range and Velocity Accuracy from Radar Measurements," *MIT Lincoln Labs Report*, pp 312-326, Feb. 1955.

[11] R. Manasse, "Summary of Maximum Theoretical Accuracy of Radar Measurements," *MITRE Tech Series Report 2*, April 1960.

[12] P.J. Kahrilas, "Design of Electronic Scanning Radar Systems," *Proc. IEEE*, pp 1763-1771, Nov. 1968.

[13] P.J. Kahrilas, "HAPDAR," *Proc. IEEE*, pp 1967-1975, Nov. 1968.

[14] L.A. Rondinelli, "Effects of Random Errors on the Performance of Antenna Arrays of Many Elements," *IRE National Conv. Record*, Part 1, pp 174-189, 1959.

[15] J.L. Allen, "The Theory of Array Antennas," *MIT Lincoln Labs. Tech. Report No. 323*, July 1963.

[16] J. Frank & J. Ruze, "Steering Increments for Antisymmetrically Phased Arrays," *IEEE Trans.* Vol. AP-15, pp 820-821, Nov. 1967.

[17] P.J. Kahrilas — unpublished notes.

[18] D.M. Jahn, "Smoothing of Sampled Data," *Sperry Rand Internal Report No. 5284.7345*, Feb. 4, 1958.

[19] R.M. Scudder and W.H. Sheppard, "AN/SPY-1 Phased Array Antenna," *Microwave Journal*, p 55, May 1974.

[20] T.C. Cheston and J. Frank, "Array Antennas," Chapter 11 in [2].

List of Symbols

A	= amplitude — volts
$\dot{A}z$	= azimuth angle rate of target — radians/second
E	= ray elevation angle at surface — degrees
$\dot{E}$	= elevation angle rate of target — radians/second
$\sqrt{G}$	= antenna voltage gain
G_1	= precomparator amplitude error — gain ratio
G_2	= post comparator amplitude error — gain ratio
G_{SL_m}	= maximum sidelobe level of antenna pattern
$H'^2(f)$	= differentiator function
K_M	= normalized monopulse slope
K_R	= 1 to 2 (1.5 typical) resolution factor
L	= length of tropospheric portion of path — meters (below 5 Km altitude)
L_R	= radial target span — meters (radial distance between major scatterers)
L_x	= cross range target span — meters (cross range distance between major scatterers)
LSB	= least significant bit
N_e	= total number of elements
N_s	= surface refractivity

N_t = tropospheric refractivity in vicinity of target
P = number of phase shifter bits
Q = number of computer bits
R = range to target — meters
R_L = maximum residue level of signal waveform
R_M = maximum instrumented range — meters
S_o $= \dfrac{\theta_o}{K_M \sqrt{G}}$ = reciprocal of difference slope — milliradians/gain differential

S/N = single-hit signal-to-noise ratio
$(S/N)_o$ = single pulse detectability factor
T_o = time between looks — seconds or look time on target
T_d = signal time duration — seconds
T_f = finite time on target — seconds
or finite length of pulse
or finite length of pulse train
T_s = sample period — second
V_M = maximum target velocity — meters/second
V_T = target radial velocity — meters/second
V_a = average wind speed — meters/second
V_d = drift velocity of air mass — meters/second
V_m = maximum instrumented velocity — meters/second
W = aperture width — meters
W(f) = refractivity spectrum for average conditions
a = maximum acceleration — meters/second^2
b = baseline length — meters
c = 3×10^8 meters/second — velocity of light
f_o = transmission frequency — hertz
f_c = clock frequency — hertz
f_d = doppler frequency — hertz
$\overline{f_d}$ = smoothed doppler velocity $\sim$ 1 hertz
h = radar antenna height — meters
n = number of hits smoothed
t_o = observation time — seconds
ΔE = elevation bias error — milliradians

ΔR	= range resolution — meters
ΔR_b	= range bias error — meters
Δf	= signal bandwidth — hertz (FM deviation during time T_f)
Δf_o	= half power spectral line width — hertz
Δt_s	= maximum pulse jitter between samples — seconds
Δv	= doppler velocity resolution — meters/second
$\Delta\theta$	= angle resolution — meters/second
α_{CD}	= position smoothing constant for critically damped system
α_1	= precomparator phase error — degrees
α_2	= post comparator phase error — degrees
α_3	= post post comparator (limiter) phase error — degrees
β_{CD}	= velocity smoothing constant for critically damped system
ϵ	= rss (root sum square) of phase errors — radians
θ_o	= 3 dB beamwidth — meters/second
λ	= transmission wavelength — meters
ρ	= surface reflectivity
σ_A	= amplitude error — volts
σ_{Az}	= unknown or uncorrected portion of tropospheric bias error in azimuth — radians
σ_{DL_A}	= angle dynamic lag error — radians
σ_{DL_R}	= range dynamic lag error — meters
σ_{DL_v}	= velocity dynamic lag error — meters/seconds
σ_E	= unknown or uncorrected portion of tropospheric bias error ΔE — radians
σ_{E_o}	= elevation error from baseline system — milliradians
σ_{FM}	= range FM error — meters (for linear RM with rectangular envelope waveform)
σ_{G_A}	= angle glint error — radians
σ_{G_R}	= range glint error — meters
σ_{G_V}	= velocity glint error — meters/second
σ_I	= angle instrumentation error — meters/second
σ_{M_A}	= angle multipath error — meters/second
σ_{M_R}	= range multipath error — meters

σ_{M_V} = velocity multipath error — meters/second
σ_N = velocity propagation error — meters/second
σ_{Q_A} = angle quantization error — radians
σ_{Q_R} = range quantization error — meters
σ_{Q_V} = velocity quantization error — meters/second
σ_{T_A} = angle thermal error — meters/second
σ_{T_R} = range thermal error — meters/second
σ_{T_V} = velocity thermal error — meters/second
σ_{TF} = range rate propagation error — meters/second
σ_{VAz} = azimuth coordinate velocity propagation error — meters/second
σ_{VE} = elevation coordinate velocity propagation error — meters/second
σ_b = beam pointing error — millisine radians
σ_c = range clock error — meters
σ_r = range fluctuation for average conditions — meters
σ_{rc} = range fluctuation after refractivity — meters
σ_{ro} = apparent range fluctuation observed in restricted interval — meters
σ_{p_R} = range pulse jitter error — meters
σ_{p_V} = velocity pulse jitter error — meters/second
σ_Δ = maximum dynamic range of target size
σ_θ = angle propagation error under average conditions — milliradians
σ_ϕ = phase error — radians
$\sigma_{\phi Q}$ = phase quantization beam pointing error — meters/second
τ_e = equivalent pulse length — seconds

$$(\tau_e = \frac{1}{B})$$

$\ddot{\phi}$ = maximum angular acceleration — radians/second^2
ω = target rotation rate — radians/second

Parameter Management

Chapter Ten

10.1 Definition — Parameter management is a method of obtaining maximum performance for minimum power-aperture and design complexity in an ESRS, by judicious manipulation of radar parameters.

The flexibility of a computer controlled ESRS provides the opportunity to obtain considerable economy in energy/power and time with no sacrifice in performance.

Parameter management is accomplished by varying: beamwidth, frequency, frame time/data rate, p.r.f./sampling rate, and energy/waveform as functions of operational mode, range and angular beam position.

The method is demonstrated in Chapter 14 and the results are summarized in a parameter management chart (refer to Table 14.1 — Parameter Management Chart; Chapter 14).

10.2 Beamwidth — Beam generation is one of the most important design considerations in an ESRS. It affects the basic architecture of the entire radar system.

All the options discussed are influenced by system requirements and employ different tradeoffs of: power-aperture, energy, time, waveforms, resolution/bandwidth, accuracy, efficiency, design complexity and cost-effectiveness.

Design options are presented in decreasing order of complexity and performance.

1. Transmit/Receive — simultaneous multiple pencil beams covering the full search volume.

   Each beam at a different frequency.

   Maximum power — minimum time

   In limit, the search frame time requires one echo interval and simultaneous multiple target track requires a second echo interval.

   On a time-shared basis, the search frame time is equal to two echo intervals and the sampling interval on all tracked targets is also equal to two echo intervals.

   This design requires the highest power, multiple receivers and signal processors (one monopulse set for each search beam), multiple A/D converters, complex phasing networks and displays, and high speed operation and data processing (large capacity computer).

2. Transmit — single broadened (defocused) beam that is step scanned through the search volume.

   Receive — simultaneous multiple full gain narrow beams to fill volume covered by transmit beam.

   This implementation eliminates complex transmit networks but requires high transmit energy (peak power x pulse duration) and low receive sidelobes.

   Narrow full gain transmit/receive beams are used in track.

3. Transmit — "machine gun" — a group of full gain narrow beams in rapid succession, possibly employing a different frequency for each beam.

   Receive — simultaneous multiple full gain narrow beams corresponding to transmit beam positions.

	This design reduces the energy per transmit pulse but requires adequate minimum range and a high duty cycle transmitter capability.
4. Transmit/Receive—	single broadened (defocused) beam that is step scanned through the search volume to save search time and narrow (focused) beam in track mode to improve accuracy and conserve energy and power.
	May also vary beamwidth as function of elevation angle and/or as function of echo sweep time (to maintain a constant space cell).
5. Transmit/Receive —	single full-gain beam, time-shared between search and track.

10.3 Time — Beamwidth management discussed in the previous section also included certain forms of time management. Additional forms of time management are discussed in this section.

1. Search — In many radar applications, the objects to be detected have a maximum altitude at which they can operate. Time management can be employed to decrease the interpulse period as the beam is stepped in elevation so that the echo time interval corresponds to the maximum range for each elevation beam position. For a specified maximum altitude limit, the maximum range is determined from a radar refraction chart using the lower 3 dB angular position of the elevation beam.

Another form of time management is to employ different frame times for different portions of the search volume. One method of implementation is to sweep azimuth rasters at low elevation angles twice as often as the upper ones. A possible alternative is to employ a long range and short range echo interval (and waveform) consistent with the requirements of detecting near-in pop-up targets.

2. Track — Time management of the sampling interval of tracked targets is a function of target position (R, α, β), track initiation and target dynamics (see Chapter 8).

Target Position —	A set of recursive equations to generate a sampled data servo loop in the computer. The equations are identical for range and the two angular coordinates, and differ only in the values of the smoothing constants. The sampling rate and bandwidth (smoothing constants) can be decreased continuously or in discrete steps as the range decreases and/or the angular rates increase.
Track Initiation —	During track initiation, a larger bandwidth and faster sampling rate are required than under steady state conditions to maintain tenacity of track. The bandwidth (smoothing constants) and sampling rate can be changed in preprogrammed discrete steps as a function of time after target acquisition or changed continuously as a function of the error signal in each coordinate.
Target Dynamics —	To improve accuracy and maintain dynamic performance adaptive smoothing rules can be incorporated to vary smoothing constants and sampling rate to match the current target acceleration.
	Power/energy and sampling rate requirements can be reduced by the use of higher type or Kalman filter tracking loops in exchange for increased complexity.

10.4 Energy — Energy (peak power x pulse duration) management is achieved by adjustment of peak power and/or pulse and pulse burst duration. Since transmitters are usually peak power limited, peak power control may be obtained by turning off the output tube and allowing the driver stage power to pass through the output tube to the antenna. The discrete step power reduction obtained is equal to the gain of the output tube plus the passive losses through the output tube. This method is feasible only when the output tube (cross-field amplifier for example) provides passive coupling when it is turned off. Pulse duration adjustment is employed more extensively because it provides a wider range of control for energy management.

1. Search — As previously mentioned in the discussion of time management, there are many radar applications where the objects to be detected have a maximum altitude operating limit. Energy management can be employed to reduce the transmitted energy as the beam is stepped in elevation consistent with the decrease in maximum range as determined from a radar refraction chart using the lower 3 dB angular position of the elevation beam.

The energy reduction is determined by the following formula:

$$\epsilon_{R_{n_{dB}}} = \left(\frac{R_i}{R_n}\right)^4_{dB} + \Delta_{ATD_{dB}} \qquad (10.1)$$

where:

ϵ_{R_n} = energy reduction at n^{th} elevation beam position

R_i = initial range where process begins

R_n = maximum range at n^{th} elevation beam position

Δ_{ATD} = decrease in atmospheric loss between ranges R_i and R_n

Energy management is normally determined relative to the lowest elevation beam position. An important exception may arise where a specified cumulative probability of detection is required for targets of different cross sections (aircraft and missiles) at different detection ranges. The energy reduction in these applications may be delayed several elevation beam positions to be consistent with the requirements.

2. Track — Energy management can be employed to reduce the transmitted energy as the range of each tracked target decreases relative to maximum target acquisition range. Expression (10.1) is still applicable with the following redefinition of R_i and R_n:

R_i = maximum target acquisition range

R_n = next predicted position of tracked target

The Δ_{ATD} term may be dropped for simpler implementation

3. MTI — Moving Target Indicator — In many applications the specified detection range of the radar is much longer than the maximum clutter range. Sea clutter only extends several miles and distributed land clutter extends to a maximum range of approximately 100 km (54 nm). Mountainous regions may contribute discrete clutter returns at longer ranges. Consequently, energy management is used to reduce appreciably the MTI energy requirements in long range radar system applications. Expression (10.1) applies for MTI with the following redefinition of R_i and R_n.

R_i = maximum target detection range

R_n = maximum clutter range

4. Beam Scanning — In determining the power/energy requirements of an ESRS, a loss is included in the radar range equation to account for the drop-off in antenna gain as the beam is scanned off broadside. This round trip loss is approximated by $\sec^3 \xi$ where ξ is the angle from array normal. The round trip aperture loss varies as $\sec^2 \xi$. The secant function is raised to the third power to include the round trip effect of antenna element mismatch. For a scan angle off broadside of 60° this loss is approximately 9 dB. It is therefore highly desirable to employ energy management as a function of scan angle to conserve energy in all modes of operation. The energy can be decreased continuously as ξ decreases or, more practically, it can be controlled in discrete binary increments. A simple computer algorithm computes ξ for every scan angle and determines the energy management for search, track and MTI operational modes.

An alternate compensation for the $\sec^3 \xi$ loss is to decrease the spacing between beam positions as the beam is scanned away from array normal. This can be accomplished automatically by simply stepping the beam by a fixed milliradian (not millisine radian) increment.* The fixed increment can be equal to the 3 dB beamwidth at array normal. As the beam widens with scan angle the overlap between beam positions increases thus reducing the beam shape loss and increasing the probability of detecting the target in more than one beam position.

10.5 Waveform — Waveform management is feasible with either digital or analog matched filter implementation. In the most common case of peak power limited transmitters, the pulse and/or pulse burst duration is varied as a linear function of energy. Energy management is achieved by pulse length management. Both the pulse-length and the time-bandwidth product are adjusted in binary increments to maintain

*Suggested by D.K. Barton

the same compressed pulse width. Keeping the same range resolution (bandwidth) for all operational modes has several advantages. No time is lost nor implementation required to perform a range cell search in going from normal or MTI search to track acquisition. There is less likelihood of multiple target ambiguity in handover from one operational sector to another. The receiver bandwidth is the same for all waveforms and operational modes thus simplifying implementation and threshold level setting. Binary incremental control of the pulse length and time-bandwidth product simplifies the design of the signal processor. In an analog or digital signal processor the various waveforms are obtained by switching in-or-out binary segments of the matched filter network. To avoid excessive design complexity the number of waveforms employed should be limited to approximately ten.

An exception arises to maintaining the same resolution for all waveforms when the energy required for the search function is much higher than required for the track function and the time-bandwidth product is too large to permit a practical matched-filter processor design. Another exception may arise when a larger bandwidth is required for MTI operation to obtain the required clutter reduction.

Waveform management is employed in search, track and MTI operational modes as a function of range and beam scanning angle in a similar manner to energy management. A lower limit of time-bandwidth product is approximately 25. Below this value it becomes too difficult to achieve acceptable residue levels. In many MTI applications it is possible and desirable to use a single frequency waveform with a pulse width equal to the compressed pulse width of the coded waveforms.

10.6 Power — In an ESRS design employing parameter management the average power transmitted in each operational mode is determined by dividing the total energy transmitted by the corresponding total time interval. To assist in this calculation it is recommended a parameter management chart be prepared similar to Table 14.1 in Chapter 14. An appropriate reduction factor is required in the calculation to take into account the energy conserved as a function of beam scanning angle (refer to Chapter 14).

10.7 Time and Power Budgets — The time and power budgets of an ESRS indicate the effect of time sharing of operational modes on time and power allocations. They are usually expressed as fractions or percentages of total time and total average power (refer to Chapter 14). These budgets are used to size a system and to select suitable subsystem components. Their evaluation can critically affect the system architecture. In the extreme, they can influence the selection between a single beam and multibeam ESR design. For best results, proper management of time and power budgets must be individually tailored for each ESRS application to obtain a balanced design.

10.8 Duty Cycle — The ratio peak power to average power or pulse duration to interpulse period is called the duty cycle of the radar. In an ESRS that has multimode operation on a time-shared basis there are different duty cycles for each operational mode. In addition, the short-term duty cycle is different from the long-term duty cycle. Short-term may range from one or two milliseconds (one interpulse period) to approximately 100 milliseconds (computer major cycle time).

Duty cycle management is required to meet the short-term and long-term duty cycle transmitter specifications. It is accomplished by determining all the short-term duty cycles for the various operational modes and then developing an algorithm to obtain an appropriate mixture of search, track and MTI scans per computer major cycle consistent with the transmitter specifications. A second algorithm is developed to continually monitor the long and short term duty cycles of the transmitter and not allow them to exceed acceptable design limits.

10.9 Frequency — Pulse-to-pulse frequency agility conserves energy/power in applications where a high probability of detection is desired on noncoherently integrated target returns for Rayleigh and one-dominant-plus-Rayleigh target models. Swerling cases II and IV involve pulse-to-pulse decorrelation of echo amplitudes (achieved by use of pulse-to-pulse frequency agility) and require less signal-to-noise ratio for a specified n and probability of detection ($\geqslant 0.5$) than Swerling cases I and III that involve scan-to-scan decorrelation but are relatively constant (slow fluctuating) pulse-to-pulse. Frequency agility also economizes on energy/power required for cumulative probability of detection if the individual probabilities of detection are significantly larger than 0.34.

Another use of frequency management, discussed in Section 10.2, is for beam position identification when multiple receive beams are used.

In addition to the other advantages of frequency management, frequency agility provides an effective electronic counter-countermeasure capability.

List of Symbols

R_i	= initial range where process begins
	or
	= maximum target acquisition range
	or
	= maximum target detection range
R_n	= maximum range at n^{th} elevation beam position
	or
	= next predicted position of tracked target
	or
	= maximum clutter range
Δ_{ATD}	= decrease in atmospheric loss between ranges R_i and R_n
ϵ_{R_n}	= energy reduction at n^{th} elevation beam position
	or
	= energy reduction at next predicted position of tracked target
	or
	= energy reduction at maximum clutter range

ESRS Parameters

Chapter Eleven

The remaining elements required as background to the use of a system design methodology (Chapter 12) are contained in this chapter.

11.1 Operational Scenario — An operational scenario is an overview of the complete automatic (and manual, if necessary or requested) operation of an ESRS during a mission. It exemplifies the sequence of events occurring under normal operating conditions. The scenario selected should be intentionally the most complex required in a mission to demonstrate maximum use of the ESRS resources.

An operational scenario serves the following functions:

1. Demonstrates that the ESRS has the capacity to meet all the performance requirements.
2. Forms a basis for the development of a computer software operational program by defining radar actions/modes and radar scheduling/templates.
3. Forms the basis for the development of system power and time budgets.
4. Forms the basis for the development of a system functional block diagram.
5. Contributes to the development of system error and loss budgets.

6. Contributes to the development of a system parameter listing.

The information contained in an operational scenario is outlined for a typical ESRS application.

ESRS Missions

- I Intelligence
- II Early Warning
- III Space Track

Operator directed missions and override options are discussed at the end of each sequence.

Mission Classification Operational Modes

1. Long term search fence
   a. Coverage volume
   b. Frame time
2. Verification
   a. When performed on detected objects
3. Correlation
   a. When performed on existing track files to eliminate tracking the same target more than once
4. Track
   a. Track capacity
   b. Type of sampled data servo loop
   c. Sampling rate at track initiation
   d. Time duration for track initiation
   e. Steady-state sampling rate
   f. Time duration for steady-state track to determine trajectory to required accuracy
   g. Reacquisition algorithm for lost targets

Each mission selected from above data is discussed separately.

I **Intelligence** — Threat model description including likely trajectory from launch point to impact point.

1. Pre-defined Mission Event Profile Selection Based on Data Obtained

   or

2. Secondary Search
   - a. Specify volumetric coverage around detected objects
   - b. Drop long term search fence at this time
3. Automatic Object Classification
   - a. Method
   - b. Number of objects - capacity
   - c. Time duration
4. Metric Data
   - a. Accuracy
   - b. Sampling rate
   - c. Time duration
5. Signature Data
   - a. Types
   - b. Sampling rate
   - c. Time duration
6. Raster Scan Search
   - a. Activated when radar resources become available
   - b. Specify coverage and frame time
7. Mode Mixes
   - a. Method of time-sharing
   - b. Parameter management
8. Operator Controls
   - a. Start/stop system operation
   - b. Select mission
   - c. Add, eliminate or modify all modes
   - d. Displays
   - e. Print Hard Copy

II **Early Warning**
(Detailed as for I)

III Space Track
(Detailed as for I)

It sometimes is useful to include an operational flow diagram in a scenario description to succinctly outline the method (or algorithms) employed for mode selection and radar scheduling.

A mission scenario chart is another valuable tool for presenting a succinct summary of the real time activities during a mission and the corresponding radar parameters for each activity. A typical mission scenario chart contains a list of activities (search fence, tank track, tank sizing, secondary search, fragment track, fragment sizing, reentry vehicle track, etc.) in the first column in the real time sequence that they occur. In each row of the chart the following information is provided for each activity: time duration; range interval; coverage volume; frame time; number of objects; sampling rate on each object; target size(s); signal-to-noise ratio; false alarm rate; waveform; average power; dwell time or pulse repetition frequency; time usage (fraction of time-sharing); and duty cycle.

11.2 Time and Power Budgets — Time usage for each operational mode in a time-shared system is defined as the fraction of time devoted to that function. The time budget consists of a summary table of the operational modes and their corresponding time usage. The total time usage for all operational modes cannot exceed unity for a single beam ESRS design. The total time usage may be minimized by the use of "machine gunning," beam widening (defocusing) in search, larger beam position stepping increments in search or time management (refer to Chapter 10). If none of these techniques reduce the total time usage to unity then it is necessary to use a multiple simultaneous beam ESR design.

A simple example will illustrate the use of machine gunning to reduce time usage.

Requirement: *Satellite Detection*

Coverage — 300-500 n mi / 500 beam positions

Frame Time — 5 s

Track — 100 objects

Sampling rate — 1/s per target

Retrace time — 150 μs

Pulse length — 200 μs

Peak power — 10 Mw

$$T_u = \frac{\sum_{i=1}^{N} T_{Di}}{T_T} \tag{11.1}$$

Where

T_u = time usage

N = total number of beam position or targets tracked

T_{Di} = dwell time of ith beam position or i^{th} target tracked

T_T = search frame time or sampling interval of tracked target

For search

$$T_u = \frac{500\ (500\ \text{nm} \times 12.36\ \mu\text{s/nm} + 150\ \mu\text{s})\ 10^{-6}}{5\ \text{s}}$$

$$T_u = .633$$

For track

$$T_u = \frac{100\ (500\ \text{nm} \times 12.36\ \mu\text{s/nm} + 150\ \mu\text{s})\ 10^{-6}}{1\ \text{s}}$$

$$T_u = .633$$

Time Budget

Mode	T_u
Search	.633
Track	.633
Total	1.266 (> 1) NG

Now the radar template for machine gunning in pairs is:

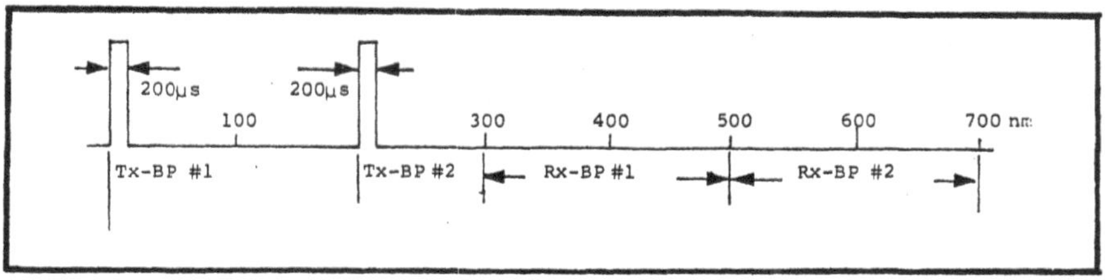

∴ For search

$$T_u = \frac{250\,(700\text{ nm} \times 12.36\ \mu s/\text{nm} + 150\ \mu s)\,10^{-6}}{5\text{ s}}$$

$$T_u = .4401$$

For track

$$T_u = \frac{50\,(700\text{ nm} \times 12.36\ \mu s/\text{nm} + 150\ \mu s)\,10^{-6}}{1\text{ s}}$$

$$T_u = .4401$$

Time Budget

Mode	T_u
Search	.4401
Track	.4401
Total	.8802 (< 1) OK

The power budget consists of a summary table of the operational modes and their corresponding average power usage.

For the previous example:

Search

$$P_{AV} = \frac{500 \times 10\text{ Mw} \times 200\ \mu s}{5\text{ s}} = 200\text{ Kw}$$

Track

$$P_{AV} = \frac{100 \times 10\text{ Mw} \times 200\ \mu s}{1\text{ s}} = 200\text{ Kw}$$

∴ Time and Power Budget

Mode	Time %	Power Kw	or	%
Search	44.01	200		50
Track	44.01	200		50

These results indicate that the full time usage is less than unity. Therefore, the search coverage sector and/or the target track capacity can be increased. Alternately, the frame time and/or track sampling interval can be decreased. Both options increase the average power usage.

11.3 Functional Block Diagram — The functional block diagram contains the major subsystems of the ESRS. It is based on the ESRS design as evolved from the operational scenario and the system design methodology process. The functional block diagram serves as a road map for describing the operation of the hardware and a guide for the allottment of error and loss budgets. It is also the starting point for a development of a full scale schematic diagram.

A generalized ESRS functional block diagram is shown in Fig. 11.1. The main subsystems are:

1. Array antenna
2. Beam steering unit (special purpose computer)
3. Transmitter
4. Receiver
5. Exciter
6. Signal processor
7. Computer
8. Radar Interface unit
9. Control console and displays
10. Peripheral units (magnet tapes, discs, line printer)
11. External interface

An illustrative ESRS is described. In a typical operation, target designation data may be supplied to and taken from the radar installation through the external interface equipment. In an experimental or instrumentation application a time reference is supplied for data recording from a base standard. The interface equipment uses this data to generate an input to the computer, representing the firing time of the transmitter. The computer references the designated coordinates to the radar site location and then transforms them to sine space array coordinates. The α, β coordinates are supplied to

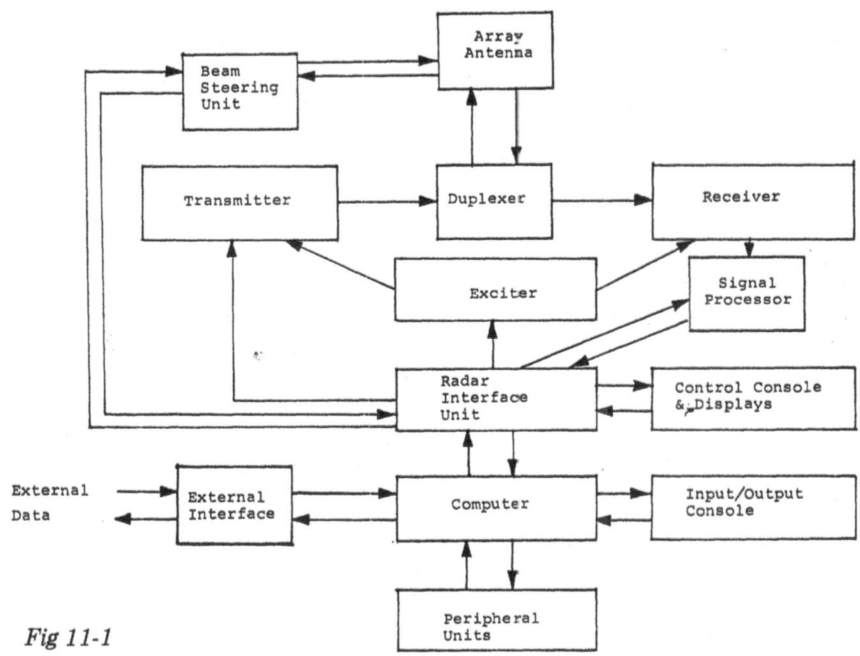

Fig 11-1

ESRS Functional Block Diagram

the beam steering unit via the radar interface unit. The beam steering unit generates and sets the appropriate phase in each active element of the array antenna to form a beam in the designated direction. Concurrently the appropriate waveform is generated and mixed or multiplied to the transmission frequency in the exciter prior to amplification in the transmitter. On completion of these events the synchronizer in the radar interface unit sends a trigger to the transmitter causing it to fire a burst of energy through the array in the designated direction. A computer programmed search function is performed in the designated sector. Target amplitude, range and angular error signals are obtained at the output of the receiver in analog form. Analog or digital matched filter processing is performed on the signal waveform in the signal processor prior to transformation to a digital format for entry into the computer via the radar interface unit. The track acquisition mode starts when a signal exceeds the threshold level. The computer performs the necessary data smoothing for generation of best estimates of position and velocity (and acceleration). The tracking and AGC operations are performed by a set of recursive equations that generate a sampled data servo loop to the computer. Type II, III or Kalman filter designs are employed. The targets are tracked in range and α, β sine space coordinates. The search and track functions are continued on a simultaneous or time-shared basis, depending on the ESRS design. The system automatically returns to a search-acquisition mode when a target is lost. Coast and manual modes are available.

The control console contains displays to observe the search sector and to monitor the tracked targets. Operational, test and calibration modes are selected at the console. The built-in test equipment includes a monitoring system using a frequency coded RF pilot pulse to equalize the three monopulse channels in both gain and phase. The computer checks itself with self-diagnostic test routines, performs go-no-go tests on all the subsystems and components in real time and off line, and the status is reported on a paper printout. The printout includes the number and location of malfunctioning phase shifters in the array.

The mission data output of the ESRS is sent back to a control center through the external interface unit and a data modem, and/or placed on magnetic tape for further analysis.

Table 11.1

ESRS Losses

	Loss Source		Equation and/or Contribution	Typical Magnitudes (dB)
1.	Atmospheric 2 way	(1)	1. Tropospheric absorption of radio waves (100 MHz to 100 GHz) by the oxygen and water vapor of the earth's atmosphere	0 to 6
		(2)	2. Weather (clouds, rain, fog, hail, snow)	0 to 10
		(1)	3. (T_s) Sky noise temperature due to atmospheric absorption and galactic sources (0 to 300°K)	
2.	Beam scanning 2 way		$L_{BSC} = \sec^3 \xi$ (11.2) L_{BSC} = beam scanning loss — 2 way ξ = scan angle from array normal-degrees	0 to 9
3.	Beam shape 2 way		Range equation derivation is based on constant gain (rectangular beam shape) and requires a loss factor to compensate for gain drop off (usually to 3 dB edges of beam pattern) of actual beam shape (sin x/x or gaussian)	0.8 to 3.2
		(3)	$L_{BS} \approx$ 1.6 dB one dimension, 3.2 dB two dimension; L_{BS} = beam shape loss	

4.	Array Antenna		3 to 6
	1) Radome	Dielectric (T_s)	0.1 to 1.0
	2) Front element	mismatch (T_s) insertion (T_s)	0.05 to 0.15 0.05 to 0.10
	3) Phase shifter	insertion (T_s)	0.5 to 1.5
	4) Rear element	mismatch — space feed array (if used) (T_s) insertion (T_s)	0.05 to 0.10 0.05 to 0.10
	5) Spillover	Space feed array (if used)	0.7 to 1.8
	6) Corporate feed	consists of waveguide, cable or strip line (T_s)	0.4 to 2.0
	7) Similarity	Loss due to amplitude and phase errors — manufacturing tolerances of array, elements, feeds, phase shifters and temperature distribution	0.2 to 0.5
	(4)	$L_s \approx 10 \log (1 + \sigma^2)$ (11.3) L_s = similarity loss σ^2 = rss of amplitude and phase errors	
	8) Obliquity	Collimation phase and amplitude losses from feedhorn to collector elements of a space feed array (if used)	0.1 to 0.2
	9) Phase Shifter Quantization (5)	$L_\phi = \frac{\pi^2}{3 \cdot 2^{2P}}$ (11.4) L_ϕ = phase shifter quantization loss P = number of phase shifter bits	.06 to 1.0
	10) Illumination Taper	Aperture distribution that produces a radiation pattern for the specified beamwidth and side lobe level	0 to 1.5
	11) Feedhorn	Space feed array feedhorn (simple or monopulse) (if used) (T_s)	0.2 to 0.5
5.	Duplexer/ Circulator	(T_s)	0.2 to 0.5
6.	Rotary joint (if used)	(T_s)	0.1 to 0.5
7.	Waveguide Cables & Connectors	(T_s)	0.1 to 3.0 1.0 to 6.0
8.	Pressure window, isolator & harmonic filter	(T_s)	0.1 to 1.5 0.1 to 1.5 0.1 to 1.5
9.	Receiver protector	T-R switch and/or limiting diodes – (T_s)	0.2 to 1.0
10.	Preamplifier	Parametric amplifier (cooled or uncolled), tunnel diode, transistor or TWT (T_s) Noise Figure Gain	 1.0 to 6.0 10 to 30
11.	Preamplifier diode switch-before (if used)	Increased radar system dynamic range by switching out preamp for strong signals → (T_s)	0.2 to 1.0
12.	Preamplifier diode switch-after	(T_s)	0.5 to 2.0
13.	RF Bandpass Filter	(T_s)	0.5 to 2.0
14.	RF Mixer	(T_s) Noise Figure conversion gain/loss	6 to 8 -6
15.	Matched Filter	Mismatch loss of signal waveform decoding and weighting filters	0.4 to 2.0
16.	CFAR	Constant False Alarm Rate circuit loss	1.0
17.	Range Gate Width (tracking)		0.5 to 1.5
18.	System Degradation	System degradation margin allotted for field operational degradation between maintenance intervals	1.0 to 6.0

11.4 Error and Loss Budgets — A well designed ESRS has top-down traceability through:

Threat model

System performance requirements

System radar parameters
(including error and loss budgets)

Subsystem performance requirements

Subsystem parameters
(including error and loss budgets)

The system error budget is readily determined by use of tables 9.1 through 9.3 of Chapter 9.

The estimation formulas and/or contributions to the ESRS loss budget are summarized in Table 11.1. In some cases the formulas have been modified slightly to generalize their usage. Typical magnitudes of the loss contributions are provided in the last column of the table to guide the ESRS designer in the allocation of loss budgets. The reference sources are indicated in parentheses. All the receive losses labelled (T_s) are ohmic and are included in the system noise temperature determination (see Chapter 1) in the use of the radar equation.

11.5 Parameter Listing — An ESRS parameter listing provides a concise overview of the key system characteristics and design features. Supplementary detail can be made available but the listing itself should be limited to a compact one page summary for easy reference. A representative listing is contained in Table 11.2 and exemplified at the end of Chapter 14.

Table 11.2

ESRS System Parameters

1. Radar type	Number of radars and/or arrays Single beam or multiple simultaneous beams One or two dimensional electronic scan Space or confined feed Tilt angle and maximum scan angle
2. Frequency	Center transmission frequency Tuning range Instantaneous bandwidth
3. Power	Transmitter peak power Transmitter average power Transmitter duty cycle Prime power
4. Coverage	Target fluctuation models Probability of detection False alarm rate Target radar cross sections Range Elevation angle Azimuth angle Altitude
5. Search frame time(s)	For each search sector With and without time-sharing other radar modes
6. Track Capacity	Number of each target type and corresponding sampling rate

7. Antenna	Type (space or confined feed) Simple or monopulse (amplitude or phase) Thinned (amount) or fully-filled Number of active elements Number of dummy elements Size Beamwidth Directive gain (area gain) Losses (not itemized) Type of illumination taper Maximum sidelobe level RMS sidelobe level Null depth Polarization
8. Phase Shifters	Type (diode, ferrite or garnet) Analog or digital Number of bits Insertion loss RMS phase error RMS amplitude error
9. System Noise Temperature	Value Sky temperature used Type of preamplifier (parametric - cooled or uncooled, tunnel diode, transistor or TWT) Noise figure and gain of preamplifier
10. Waveform(s)	Type(s) (pulse duration and coding)

11. MTI	Type(s) (pulse doppler or number of cancellers and stagger) Improvement factor Coverage, frame time, when used
12. Interpulse period(s)	Itemized or indicate range
13. Resolution	Range Angle(s) Radial velocity (doppler)
14. Accuracy	Range Angle(s) Radial velocity (doppler) Amplitude (cross section)
15. Beam switching	Phase shifter setting and settling time
16. Beam computation	Time between early trigger and main trigger (retrace time)
17. Transmitter Interpulse Noise	Referenced to one milliwatt per hertz
18. Computer	Type - manufacturer & model Speed Storage Word length Number of input/output channels
19. Mean time between failures	

20 Mean down time	
21. Clock	Type Frequency Stability
22. Receiver	Type (simple or monopulse - conventional or single-hit normalization) Dynamic range Pre comparator, post comparator and post post comparator (limiter) Phase and amplitude errors
23. Signal Processor	Type (analog or digital)
24. Tracking filter(s)	Range (type II, III or Kalman) Angle (type II, III or Kalman)

References

[1] L.V. Blake, "Radar/Radio Tropospheric Absorption and Noise Temperature," *NRL Report #7461*, Oct. 30, 1972.

[2] B.R. Bean, E.J. Dutton & B.D. Warner, "Weather Effects on Radar," Chapter 24 of M.I. Skolnik (Ed.), *Radar Handbook*, McGraw-Hill, 1970.

[3] L.V. Blake, "The Effective Number of Pulses per Beamwidth for a Scanning Radar," *IRE* pp 770-774, June 1953.

[4] J. Ruze, "The Effect of Aperture Errors on the Antenna Radiation Pattern," Proc. Symposium on Communication Theory and Antenna Design, *AFCRC Tech. Rpt.*, pp 57-105. ASTIA Document 117067, Jan. 1957, also in *IEEE Proc.*, April 1966.

[5] C.J. Miller, "Minimizing the Effects of Phase Quantization Errors in an Electronically Scanned Array," *Proc. 1964 Symposium*, Electronically Scanned Array Techniques and Applications, RADC-TDR-64-225, Vol. 1, pp 17-38.

List of Symbols

L_{BS} = beam shape loss — 2 way

L_{BSC} = beam scanning loss — 2 way

L_s = similarity loss

L_ϕ = phase shifter quantization loss

N = total number of beam positions

P = number of phase shifter bits

T_{Di} = dwell time of ith beam position or i^{th} target tracked

T_T = search frame time or sampling interval of tracked target

T_s = system noise temperature

T_u = time usage

ξ = scan angle from array normal — degrees

σ^2 = rss (root sum square) of amplitude and phase errors

System Design Methodology

Chapter Twelve

ESRS architect's function: *To develop radar system to meet performance requirements — preferably with simplest and least expensive design*

12.1 **Methodology** — The major ESR design considerations have been presented. The ESRS architect is now adequately prepared to appreciate and use a system design methodology based on the material contained in this and the preceding chapters.

The methodology to be presented enables the ESRS architect to evolve an integrated "balanced" system design that draws on the contributions of: performance analysts, waveform experts, array, computer, transmitter, receiver, signal processor, control consoles and displays, reliability, maintainability and quality control specialists to meet all the performance requirements at minimum: risk, complexity, development schedule and cost.

The approach is not unique nor are the results indisputably optimum.

Complexity should be looked upon with suspicion — not reverence.

The system designer has to be knowledgeable in all areas to make key decisions such as space or corporate feed or, more important, single or multiple beam design. The system architect obtains the recommendations of all the experts in their respective areas but in the final analysis he makes the critical decisions that establish the architecture of a cost-effective ESRS design to best meet system performance requirements.

Table 12.1
Relation of Threat Model/Performance Requirements to Radar Parameters

Performance Requirements	Design Considerations	Radar Parameters
Mission(s)	**Threat Scenario(s)**	**Number and types of radar modes**
Surveillance: — search mode(s) To discover/detect targets	1. Location, number and types of targets 2. Target density-resolution 3. Target dynamics 4. Reaction time 5. Environment-climate, weather, terrain	1. Volumetric coverage — maximum and minimum range, elevation and azimuth angular coverage 2. Fluctuation model(s) and radar cross section of targets 3. Waveforms for: detection, range and velocity resolution, weather and clutter rejection, signature and discrimination 4. Energy — based on probability of detection and false alarm rate for target fluctuation model(s) 5. Pulse repetition rate(s) for above range, MTI and signature requirements 6. Frame time(s) 7. Azimuth and elevation beamwidths for resolution, frame time(s), and power-aperture requirements 8. Transmission frequencies 9. Peak and average power(s)
Acquisition/Track: — track mode(s) To acquire and monitor selected targets and determine trajectories (this category can include counterweapons like artillery shells and antimissile missiles)	1. 1. to 5. Same 6. Acquisition logic — order of selection of targets to be tracked 7. Track monitoring logic — method of determining if target is still in track 8. Reacquisition logic — method of reacquiring targets that have dropped out of track	1. 1. and 2. Same 3. Waveforms for: detection, velocity and range resolution, accuracy, weather and clutter rejection, signature and discrimination 4. Energy — based on probability of detection, false alarm rate and accuracy for target fluctuation model(s) 5. Sampling rate, smoothing filter and track duration for each tracked target 6. Azimuth and elevation beamwidths for resolution, accuracy and power-aperture requirements 7. 7. and 8. Same as 8. and 9. of search mode(s)
Guidance: Uplink and downlink messages and/or homing illumination	1. Location, number and types of missiles 2. Environment — climate weather, terrain 3. Missile address 4. Kill evaluation	1. Same 2. Waveforms 3. Energy 4. Sampling rate, smoothing filter and guidance duration for each missile 5. Azimuth and elevation beamwidths for resolution, accuracy 6. 6. and 7. Same as 8. and 9. of search mode(s)

RELATION OF THREAT MODEL/PERFORMANCE REQUIREMENTS TO RADAR PARAMETERS
TABLE 12.1

The basic elements of the system design methodology are:

Transformation of threat model/performance requirements to radar parameters

Real-time sequence scenario

Performance analysis

ESRS design selection

Charts and tabulations

Iteration(s)

Each basic element of the system design methodology is described in this chapter and the application of the methodology is illustrated in Chapter 14 — Case Study.

12.2 Transformation of Threat Model/Performance Requirements to Radar Parameters — The specified threat model and/or performance requirements are analyzed and translated to radar functions and parameters. A general format for performing this operation is indicated in Table 12.1.

12.3 Real-time Sequence Scenario — The procedure for developing the real-time sequence scenario has been described and exemplified in section 11.1. In the process of developing an operational scenario to meet all the threat model and performance requirements,

1. *the number and types of radar modes are delineated.*
   The complexity and congestion of the radar modes developed by the real-time sequence of operations leads to an
2. *"initial" decision on the selection of a single or multiple beam ESR design.*
   Several scenario variations most likely will be considered before a satisfactory blend of operations is achieved.
3. *The reaction time (if not specified) is determined*
   based on the minimum time required to take counteraction. The reaction time is now budgeted into time segments for the sequential search, track and guidance functions.

12.4 Performance Analysis —

4. *An initial beamwidth selection*

and

5. *transmission frequency selection*
are made based on the requirements for the "dominant" radar mode(s).

Beamwidth selection is based on:

Resolution

Accuracy

Power — aperture

Search coverage(s) — frame time(s)

Clutter exclusion

Jammer exclusion

Frequency selection is based on:

Aperture size

Power — aperture

Transmitter(s) — power availability and cost

Microwave components - losses, availability, design complexity and cost

Detection enhancement

Horizon path loss - over smooth earth or sea

Weather - back scatter and attenuation

MTI performance - improvement factor and blind speeds

Atmospheric attenuation

Nuclear blackout

ECM (Enemy Counter Measure)

Evaluation of these factors leads to the conclusion that it is preferable to use: relatively wide beamwidths at low frequencies (300 to 2000 MHz) for search mode(s); relatively narrow beamwidths at mid frequencies (2000 to 6000 MHz) for track mode(s); and non-critical beamwidth requirements at high frequencies (4000 to 10,000 MHz) for the guidance mode(s). The overall effect on an ESR design is that the beamwidth is determined by the tracking requirements and the operating frequency is at UHF, L or S band for ESRS's employing only search and track modes; while the higher frequencies in the S or C band region are used by ESRS's employing search, track and guidance modes. If there is any difficulty meeting the search frame time and track accuracy requirements consider the options of using: defocused beam for search functions and narrow beam for track, "machine gun" transmission, multiple beam reception or any of the other options discussed in section 12.5, ESR System Design Selection.

6. *Tilt angle selection*
   (based on considerations discussed in section 7.6) is the last initial design decision required prior to performing a detailed system analysis.

The analysis of the ESRS is performed for each radar mode defined by the real-time sequence scenario. The functional block diagram is developed as the radar parameters formatted in table 12.1 are determined by the following step-by-step systematic procedure:

Search Mode(s)

1. Maximum detection range of target(s)
2. Signal bandwidth - resolution
3. False alarm number
4. Search pattern
5. Frame time
6. Cumulative probability of detection(s)
7. Detectability factor(s)
8. System design - E S technique
9. Array design
10. Loss budget
11. Energy - radar range equation
12. Peak power
13. Waveform(s)

Track Mode(s)

1. Waveform(s)
2. Signal-to-noise ratio(s) at maximum track range(s)
3. Accuracy - error budget
4. Monopulse technique
5. Array design (revise - if necessary)

6. Beamwidth (revise - if necessary)
7. Frequency (revise - if necessary)
8. Receiver
9. Signal processor
10. Tracking filter(s)
11. A/D converter(s)
12. Loss budget

Guidance Mode(s)

1. Waveform(s)
2. Signal-to-noise ratio(s) at maximum guidance range(s) (uplink/downlink messages and/or homing illumination)
3. Accuracy - error budget
4. Beam splitting technique
5. Guidance design
6. Beamwidth (revise - if necessary)
7. Frequency (revise - if necessary)
8. Transmitter (on missile)
9. Receiver (on missile and/or on ground)
10. Signal processor (on missile and/or on ground)
11. Tracking filter(s)
12. A/D converter(s)
13. Loss budget

MTI — All Modes

1. Waveform(s)
2. Signal-to-clutter (weather, chaff, etc.) ratio(s) at maximum MTI range(s)
3. Improvement factor - required
4. MTI design
5. Improvement factor - obtainable
6. Beamwidth (revise - if necessary)
7. Frequency (revise - if necessary)
8. Blind speeds
9. Rejection velocity range

Parameter Management — All Modes

1. Beamwidth
2. Beam scanning
3. Time
4. Energy
5. Waveform
6. Power
7. Time and Power budgets
8. Duty Cycle - long and short term
9. Frequency

12.5 ESR System Design Selection — The ESR system design evolves concurrently with the determination of the radar parameters, scenario and system analysis. A format (check-off list) for ESR system selection is outlined in table 12.2 and discussed below.

1. Volumetric Coverage — In applications where hemispherical coverage is required the choice exists among the following:

a. four arrays, each covering approximately 1/4 of the hemisphere, or

b. three arrays, each covering approximately 1/3 of the hemisphere, or

c. one array - 360° lens (like Luneburg lens) or

d. one array, mechanically rotating in azimuth and electronically scanning a pencil beam in elevation from 0 to 90 degrees

The choice between the use of three or four arrays depends on a cost-effectiveness trade-off among the added design complexity of additional elements, larger element mismatch and aperture losses associated with the three arrays in scanning a larger angle from array normal, to the savings of a whole array.

A similar cost-effectiveness tradeoff is required between a single lens type array increased complexity and design risks versus multiple arrays.

The single mechanically rotating array can only be used for hemispherical coverage if the data rate and performance requirements can be met, and if the target dynamics and target density are low enough to employ track-while-scan instead of continuous interpolative-null tracking.

Table 12.2

ESR System Selection

Function or Design	Choice
Volumetric Coverage	1, 2, 3 or 4 array antenna(s)
Arrays	single: transmit/receive array separate: transmit and separate: receive arrays fully filled or thinned
Beam Generation	single: time-shared beam multiple: simultaneous beams (cluster or independent)
Power Source	single: transmitter multiple: transmitters (tube or solid state)
Electronic Scanning Technique	frequency, phase, frequency/phase, real-time, or electronic feed switching
Power Feed Network	optical (space) or confined feed network
Beam Splitting	amplitude weighting, amplitude monopulse, phase monopulse amplitude/phase monopulse
Parameter Management	conventional fixed: PRF, frame time, sampling rate, tracking loop bandwidth, waveform, frequency and beamwidth variable: PRF, frame time, sampling rate, tracking loop bandwidth, waveform, frequency and beamwidth
Illumination Taper	identical taper for: transmit and receive antenna patterns different taper for: transmit and receive antenna patterns different Δ and Σ tapers for the two cases

2. *Arrays* — There is a significant economy in building space, array structure, antenna elements, phase shifters, feed network and beam steering unit in the use of a common transmit/receive array over a separate array for transmission and a separate array for reception. This economy carries with it some loss in flexibility in establishing independent receive beams.

The advantages of two arrays are that good isolation can be obtained between transmission and reception without the use of duplexers and the transmit and receive array patterns can be individually designed to satisfy the diverse requirements of trnasmission and reception. The isolation and independent pattern control can also be obtained with the use of a single transmit/receive array employing a more complex array design and having larger array losses due to duplexing and isolating equipment. Where multiple transmitters are utilized, the common array transmit/receive modules are much more complex and costly than the separate transmit modules and separate receive modules that are used in a two array design.

Two arrays are favored where the transmit and receive operations and beamwidths are quite different and long time duration waveforms are employed for transmission.

In some applications it is advantageous to utilize a divided aperture (inner core of active elements for transmit, inner and outer active elements for receive).

Another design selection is between a thinned and fully-filled array. A thinned array economizes (up to 50 percent or more) on phase shifters and phase shifter bits; plus the auxiliary drivers, power supplies, cabling, cooling and maintenance. Thinning provides the same beamwidth and maximum sidelobe level as a fully-filled array but there is a one-way gain loss directly proportional to the amount of thinning. Thinning has inherent growth potential in that the dummy elements can be replaced by active elements at a later date and provide approximately 6 dB more of round trip antenna gain. Thinning is favored in applications where the gain can be sacrificed.

3. *Beam Generation* — Because of the design simplicity, a single beam is preferred for performing all the search, multitarget track and multimissile guidance functions on a time-shared basis. In the rare case where there is inadequate real-time to: search the specified coverage volume; track the specified number of targets at a sampling rate consistent with the target dynamics and performance requirements; and perform multimissile guidance, it is necessary to generate simultaneous transmit and receive beams, or employ broadened (defocused) beams, to satisfy the multimode requirements.

4. *Power Source* — The peak and average power requirements, cost and the system reliability requirements determine whether it is feasible to employ a single transmitter, a group of transmitters or a transmitter module for each active element of the array. An additional cost-effectivetrade-off between tubes and solid state devices is required in situations where solid-state transmit modules are feasible for the applications. Again the simplest design is the use of a single transmitter provided it can meet the power and reliability requirements.

5. Electronic Scanning Technique — A frequency scanning technique is the simplest to implement but suffers from limited instantaneous bandwidth capability (approximately 5 MHz) and has the additional restriction of requiring a relatively large transmission frequency bandwidth of operation (approximately 6 percent of the carrier frequency). The former limits the transmitted waveform signal bandwidth and the latter requires a relatively broadband transmitter operating frequency range. Frequency scanning is not particularly suitable for high-precision tracking because of the variation of frequency with beam position which in turn affects the scintillation characterictics of the target echo.

A phase scanning array has the advantage over a frequency scanning array of handling larger bandwidth signals for the same size aperture, operating frequency and scan angle by an order of magnitude improvement (approximately 50 MHz). In the case where large bandwidth signals of the order of several hundred megahertz are to be processed, it is necessary for the array to employ realtime delay devices or geodesic lens concepts similar to the Luneburg lens.

It is appropriate in some applications to frequency scan in one angular dimension and phase scan in the other angular dimension.

6. Power Feed Network — The two main categories are optical (space) or confined feed networks. The optical feed has the advantage of being simple and providing monopulse beam splitting by the use of a conventional monopulse horn, whereas the equivalent of three confined feed networks are required to obtain monopulse beam splitting. The optical feed array has the disadvantage of requiring more space than the confined feed array since F/D ratios of 0.5 to 1.0 are used.

The confined feed has the advantage of providing equal power distribution to each active element in design applications where a transmitter module is used at each radiating element, with the flexibility of having intermediate power amplifiers interlaced within the feed structure to avoid use of a single high-power exciter source. In addition, it is possible to design the feed for multiple squinted beams if they are required. The array antenna gain losses are approximately the same for the optical and confined feed designs.

7. Beam Splitting Technique — The most accurate beam splitting technique with minimum sidelobe levels and ambiguities is amplitude monopulse. Phase monopulse designs produce poor sidelobe levels and ambiguities that require additional system complexity to suppress.

The simplest and least accurate beam splitting technique is to monitor the intensity of the target echoes in adjacent beam positions and determine the angular position by some form of greatest of selection process.

Amplitude monopulse is used in optical feed arrays whereas phase monopulse or a combination of amplitude and phase monopulse are used in confined feed arrays.

8. *Parameter Management* — In an ESR system design considerable economy in energy/power and time can be realized by varying: PRF, frame time, sampling rate, tracking loop bandwidth, waveform, frequency and beamwidth as functions of target position and target dynamics.

The extent to which parameter management is utilized in a particular design is a tradeoff between system complexity and cost effectiveness.

9. *Illumination Taper* — Independent taper design is sometimes employed on transmit and receive patterns. Uniform illumination (resulting in relatively poor sidelobe levels) is utilized on transmission to maximize energy/power on target. The round trip sidelobe levels are maintained low by utilizing a high illumination taper on reception.

In monopulse operation, it is also desirable to taper independently the sum and difference illuminations in order to minimize the sidelobe level response on both.

It is especially desirable to have independent transmit and receive illumination tapers in ESR systems employing active transmit modules at each radiating element because the transmitters are usually operated at saturation levels and consequently uniform illumination or space tapering is employed for transmission.

Where space taper or thinning is inconvenient, or when additional or variable minor-lobe control is required, it is preferrable to introduce amplitude taper which, in a receive antenna, can be accomplished with attenuators without loss of sensitivity provided there is adequate preamplification associated with each element.

Beam broadening can be accomplished with amplitude taper or phase spoiling (defocusing). Phase spoiling on receive requires processing compensations to avoid adverse effects on monopulse operation. Amplitude taper is not usually an option on transmit and only phase spoiling is recommended since the inefficiency and power dissipation problems associated with amplitude taper are often unacceptable.

A broadening factor of up to four-to-one, with good minor-lobe control and acceptable monopulse performance, can be achieved with phase spoiling provided there is at least a modest amount of thinning.

12.6 Charts and Tabulations — The charts and tabulations that are recommended to prepare in the process of evolving and defining an ESRS design are summarized below for convenient reference.

- Mission scenario flow diagram
- Mission scenario chart
- Radar scheduling templates
- Sine space coverage patterns
- Loss budget
- Error budget
- Time budget
- Power budget
- Functional block diagram
- Parameter management chart
- Parameter listing

12.7 Iterations — It is apparent that there is no unique approach to the design of ESR systems. Many of the steps outlined in the system design methodology are interdependent. The performance analysis of, say, the track mode(s) might lead to a design of selection of beamwidth, signal bandwidth, or power-aperture that is inconsistent with what is optimum (or required) for the previously investigated search mode(s). Several methods have been described for resolution of possible inconsistencies. Nevertheless, there can arise situations where reiteration of the entire design selection procedure becomes necessary. Possibly to the extent of requiring a new selection of electronic scanning technique or array design to meet, say, signal bandwidth or accuracy requirements.

With experience, the iterations are minimized or entirely eliminated. Gradually, the ESRS architect evolves by a logical systematic procedure the final "balanced" (integrated) design to meet all the performance requirements at minimum: risk, complexity, cost and development schedule.

Reference

[1] Kahrilas, P.J., "Design of Electronic Scanning Radar Systems," *Proc IEEE*, pp 1763-1771, November, 1968.

ESRS Integration, Test and Evaluation

Chapter Thirteen

Successful management and technical direction of a development program on a complex ESRS requires the preparation of a detailed integration, test and evaluation plan. The philosophy of use of analysis, simulation and test for demonstrating compliance to system specifications/performance requirements must be established and clearly delineated at the beginning of the program. Early decisions are required on the degree of subsystem checkout and test prior to initiating "marriage" tests between subsystems and subsequently system level integration, checkout, test and evaluation. A parallel development is required on the software that eventually leads to hardware/software integration.

It is necessary to delineate what is to be tested at the company facilities and at the field test site — category I and category II tests. Large arrays are usually fully assembled and tested for the first time at the field test site or at the site where the ESRS becomes operational, but capability is demonstrated prior to shipment by constructing and testing a section of the array at the company facilities — possibly 100 elements. Many future problems and misunderstandings between the company and the customer are avoided if the test plan preparation (customer approval cycle) and reporting philosophy are developed documented and agreed upon by the company and the customer in the early stages of the program.

In integration, test and evaluation plan are detailed to the degree where schedule, functions to be performed (with associated activity times) and what is demonstrated are evident and create confidence in the feasibility and successful accomplishment of the program. The following are the major items in the development of an ESRS integration, test and evaluation program discussed in this chapter.

Program Plan/Schedule

Major Milestones

Hardware/Software Integration and Test Plans

Software Build Plan

Performance Demonstration and Evaluation

13.1 **Program Plan/Schedule** — The program plan is developed as a bar chart that includes all the functions to be performed during the full time span of the program and relates the various activities to each other at the appropriate points in the schedule (which is shown in monthly increments as the abscissa of the chart). Each major category is represented by one or two horizontal bars and within each bar the functions performed (when begun and when completed) are itemized. The program plan contains all the detailed steps beginning with initial design and concluding with equipment demonstration and evaluation. All tasks, when they occur, the length of time they take, and how they tie in with other tasks, are indicated on the chart. For example, the completion of the sequence of events: software description specification development, debugging of software on computer, checkout against computer simulated scenario and then checkout of software at the hardware/software hybrid simulation facility — must exactly coincide with when the equipment has been delivered and installed at the test site and is ready for use in hardware/software interface tests.

The program plan provides an overview of the progressive development and technical management of the project. If PERT (Performance Evaluation and Reporting Technique) monitoring is stipulated for the project, the program plan provides the traceability, as well as the basis, for the development of the PERT charts.

The major and interim milestones (discussed in Section 13.2) are indicated by diamonds and triangles respectively at the expected date of occurrence at the top of the program plan chart.

A typical program plan contains the following major categories and subdivisions:

I Major and Interim Milestones

II Equipment and Software

- A. Radar Subsystem
  1. Transmitter
  2. Array (mechanical)
  3. Array (electrical)
  4. Beam Steering Unit
  5. Receiver
  6. Exciter
  7. Signal Processor (if analog)
- B. Processing, Control and Peripheral Subsystem
  1. Computer/Data Processor
  2. Peripherals
  3. Radar Interface Unit
  4. Signal Processor (if digital)
  5. Control Console and Displays
- C. Processing and Control Software Subsystem
  1. Simulation Software
  2. Operational Software
- D. Communication Interface and Support Subsystem
- E. External Interface Subsystem
- F. Site Preparation and Technical Facility Subsystem

III System Engineering

- A. Technical Plan
- B. Specifications
- C. Design Reviews
- D. Technical Reports
- E. Reliability, Maintainability, Availability
- F. Electromagnetic Compatibility
- G. Logistics Support
- H. Standardization
- I. Personnel Subsystem
- J. Environmental Engineering
- K. Safety
- L. Quality
- M. Test Plans and Reporting

IV Program Management
  A. Reviews and Controls
  B. Packaging, Packing and Transportation
  C. Configuration and Data Management
V Aerospace Ground Equipment (AGE)
VI Initial Operating Spares

13.2 Major Milestones — Successful technical direction of an ESRS development program requires a reliable and timely method of measuring progress. Technical performance measurement is achieved by establishing confidence milestones for each performance objective. The milestones are selected to be measurable by either test or simulation and provide a time-phased build-up of confidence that the primary system objective is achieved. They are selected to indicate major technical achievements and/or equipment development. A three year program contains a total of approximately 30 to 40 milestones. It is recommended that they be jointly developed and documented by the company and the customer.

A typical time-phased milestone list for a 3 year development program is shown below:

End of Month	Milestone
0	Contract award
1	Operational site construction contract award
2	Draft System Specification
3	Facility PDR (Preliminary Design Review)
4	Commence Operational site preparation
5	Software PDR
6	System PDR
7	Approve System Specification
8	Facility CDR (Critical Design Review) System CDR
9	Fabrication of 100 element pilot array
10	Commercial Computer received
11	Facility design completed

12	Pilot array tests completed
13	Simulation Facility operational
14	Commence Facility DT & E (Design, Test & Evaluation)
15	Start radar SW (software)
16	Display SW completed
17	Start shipments to operational site
18	Basic radar SW completed
19	Building and Array structure completed
20	Transmitter fabrication completed
21	Array microwave completed
22	Search Mode SW completed
23	Commence on site D T & E
24	Start SW verification tests
25	Facility D T & E completed
26	Production of all HW (hardware) completed
27	Shipment to Operational site completed
28	On-site installation completed
29	Radar integration completed Track Mode SW completed Start System Tests
30	Search Mode tests completed
31	Guidance Mode SW completed
32	Search and Track modes tests completed
33	Search, Track and Guidance modes tests completed Start Performance Demonstration tests
34	AGE (Aerospace Ground Equipment) acquisition completed Start Availability Demonstration tests
35	Performance Demonstration tests completed Availability Demonstration tests completed Commence O & M (Operation and Maintenance)

13.3 Hardware/Software Integration and Test Plans — A system integration and test plan for a complex ESRS requires a step-by-step detailed logical build-up; beginning with component procurement and test and concluding with the integration, test and evaluation of a complete operational system. The plan is developed and subsequently modified by hardware availability and scheduling constraints.

Development and testing of hardware and software proceed independently up to subsystem integration. Subsequently, subsystem marriage and system operational tests are performed in a simulation facility in order to achieve a high degree of confidence prior to integration at the test site. Marriage tests and full system integration tests are repeated at the test site; initially with the aid of system simulators and finally in a real world environment.

A typical step-by-step test procedure at the test site is as follows:

1. Install and checkout RIU (radar interface unit)
2. Install and checkout CU (computer unit)
3. Perform marriage tests RIU/CU
4. Install and checkout DSP (digital signal processor)
5. Perform marriage tests RIU/CU/DSP
6. Install and checkout R_x (receiver)
7. Perform marriage tests RIU/DSP/R_x/CU
8. Install and checkout T_x (transmitter)
9. Perform marriage tests RIU/DSP/R_x/T_x/CU
10. Install and checkout BSU (Beam steering unit)
11. Perform marriage tests RIU/DSP/R_x/T_x/BSU/CU
12. After completion of Antenna pattern Tests, connect phased array to system
13. Install and checkout Control Console and Displays
14. Perform marriage tests
15. Install and checkout ground Guidance equipment
16. Perform marriage tests
17. Perform ESRS Acceptance Tests

The above tests are basically functional tests of the equipment. The next series of tests are system tests as follows:

18. Checkout all search modes employing a system target simulator that provides both i.f. and r.f. signals and can radiate to the ESRS through waveguides connected to test towers.

19. Checkout the correlation and validation of detected targets and proceed to initiate and sustain track of multiple targets employing the system target simulator. Demonstrate capability of sustaining track with maximum target dynamics and passing through different sectors with corresponding waveform changes, resolution capability on multiple targets, complete range and sine space sector coverage, etc. This test includes full capability of multi-sector search, detection, correlation, validation, track acquisition, and track maintenance of multiple targets.
20. Add to the above capability the operation of the Control Console and Displays.
21. Clutter calibration — The ESRS radiates into space and clutter mapping is performed. The ΔR, $\Delta\alpha$, $\Delta\beta$ (sine space coordinates) location of clutter residues remaining after MTI cancellation are stored in the computer.

    In normal operation the computer will automatically correlate returns with the stored location of the clutter residues and not establish track on clutter residues.
22. Flight tests — Full automatic radar operation is demonstrated in this test. A controlled aircraft is flown on racetrack courses through the various search sectors to demonstrate the system capabilities. Tenacity of track (track maintenance) is demonstrated when the target maneuvers. Time-tagged target position data is recorded. In addition, all raw data flowing in the system is recorded for system diagnostic purposes. Post flight linear regression lines are plotted to indicate target trajectory and radar tracking jitter. Trajectory data is recorded on "targets of opportunity" as well.
23. Checkout all Guidance Modes with system target simulator and static test missile(s).
24. Demonstrate full automatic operation with active missile(s) and drone target(s).
25. Perform ESRS Operational Acceptance Tests.

13.4 Software "Build" Plan — For initial integration of subsystems with the computer *functional software "builds"* and accompanying test drivers are provided. These functional or basic radar builds are used in tests 1-17. Some test drivers provide responses for subsystems that are not as yet included in the marriage tests. Test drivers are eliminated as subsystems are added to the marriage tests. A functional software build is a software program that is used to test an isolated system function independently for the most part from the remainder of the system. These functional builds are essential in validating hardware/software interfaces and verifying the operational status of functional subsystems prior to their introduction to, and in parallel with the development of, the overall system. These test builds allow for repeatable results to aid in hardware/software trouble-shooting. Software builds are actually building blocks of software somewhat analogous to subsystems being building blocks of radar hardware.

A system software build is a software program that is used to add new hardware and software components to an already defined system in order to incorporate a new function within the system. These builds are used to evolve the final hardware/software system. System builds are used for tests 18-25. The software in a system build is made of operational modules; some of which have been tested in a previous build and some of which are introduced to the system for the first time. These builds are driven by the expanding operational executive without the need for special test drivers.

The following disciplined procedure is employed for successful implementation of the build plan:

1. Define and freeze system requirements.
2. Write detailed software performance requirements for each build.
3. Develop the building blocks and test software.
4. Checkout the software builds in computer simulated models.
5. Checkout the software builds in hardware simulated models.
6. Checkout the software build in hardware against live targets — fully operational.
7. Store a duplicate tape of each build.

Exercise tight change control. Do not update tape until the changes made on the software build, used for integration testing and debugging of the system, have been carefully scrutinized by a control committee to assure optimum software implementation and more importantly that the changes have not degraded subsystem and/or system performance.

13.5 Performance Demonstration and Evaluation — A logical starting point for a system performance demonstration and evaluation program is to demonstrate capability of forming suitable antenna patterns (beamwidth, monopulse slope, null depth, maximum and rms sidelobe levels) and pointing the beam to designated coordinates (beam pointing accuracy). Followed by sensitivity measurements and demonstration of other range equation parameters (antenna gain, noise figure, loss budget, etc.) including detection of stationary targets at known locations.

The gradual demonstration build up is continued through the search, clutter mapping, track and guidance operational modes. Then multimode operations on one or two targets in a clear environment. Target handover capability is demonstrated to and from another radar. Finally culminating in the demonstration of all operational modes with full target density and target dynamics and adverse environment. A two week to one month availability demonstration is usually required near the conclusion of the evaluation tests. The various steps must, of course, be consistent with the detailed test plans and milestones.

In many cases a full environment of a large number of targets and guided missile weaponry may have to be partially simulated by the use of a system target simulator to supplement the limited number of live targets available for test purposes.

The test site theodolite monitors are usually supplemented with precision reference radars for comparative performance evaluation of the ESRS tracking capability. Miss-distance-indicators are utilized in the drones for accurate kill evaluation of the guided missile weaponry. Customer personnel training is also worked into the evaluation effort of large programs. In all cases, prior agreement is reached with the customer on what constitutes satisfactory performance demonstration and evaluation — as delineated in the test plans.

Reference

[1] Kahrilas, P.J. and Brower, J.D., "SAM-D Multifunction Phased Array Radar, Integration & Test Evaluation," *18th Tri-Service Symposium Record*, June, 1972.

Case Study—
Design of
Air Defense
ESRS

Chapter Fourteen

14.1 Threat Model and Performance Requirements

The task is to design a radar to cope with the following threat model:

100 aircraft — primary mission: air superiority

OR

50 aircraft — primary mission: air-to-ground attack

+

50 missiles — aircraft weaponry

Characteristics:

	cross section	V_{max}	a_{max}	R_{max}
aircraft	1.0 m^2	MACH 2	3g	200 nmi (370.6 Km)
missiles	0.05 m^2	MACH 3	10g	100 nmi (185.3 Km)

Array orientation:	true North
Tracking coverage:	± 45° Azimuth 0°-60° Elevation
Maximum altitude:	50,000 feet (15.2 Km)
Minimum range:	10 nmi (18.5 Km)
Flight formation: (Range resolution)	30 meter radial separation
Terrain:	Forest (heavily wooded hills)

Reaction time:	8 minutes from alarm to intercept 2 minutes to fire counter weapons to 70 nmi range
Accuracy:	4 millisines (1σ) 9 meters (1σ) Based on single hit requirements at a sample rate of one per second on aircraft and two per second on missiles to determine trajectory for counter weapons fire control against aircraft and missiles

Real-Time Scenario —

1. Search mode: Maintain continual search fence with adequate coverage and frame time to detect all hostile aircraft and aircraft launched missile weaponry.
2. Track mode: Perform second look verification (false alarm rejection) on all detected targets.

   Correlate coordinates of all verified targets against coordinates of targets in computer track files.

   Initiate and maintain track on new targets up to a maximum of 100 in the track files.

Maintain the search and track modes on a time-shared basis.

(A guidance mode is not required because the counter weapons have a separate fire control system that operates on designation data from the Air Defense ESRS.)

The system requires only two radar modes — Search and Track.

The initial design selection is to use a single beam ESRS.

14.2 Search

Detection Range — Aircraft

To provide the 8 minutes own aircraft reaction time, the enemy aircraft must be detected 8 minutes before they reach zero range, assuming own aircraft are located at an airport near the radar.

$$\therefore R_{A/C} = 8 \text{ min} \times 22.1 \text{ nmi/min} = 176.8 \text{ nmi}$$

where MACH 2 = 22.1 nmi/min

Aircraft detection range $\geqslant$ 180 nmi

Detection Range — Missile

The air-to-ground missiles have an effective range of 100 nmi and travel at MACH 3 (aircraft plus missile velocity relative to ground) towards the target. They must be detected close to their maximum launch range but at least a sufficient distance to allow the 2 minutes reaction time required for the antimissile missiles.

$$\therefore R_{MSL} = 2 \text{ min} \times 33.3 \text{ nmi/min} = 66.6 \text{ nmi}$$

where MACH 3 = 33.3 nmi/min

Missile detection range $\geqslant$ 70 nmi

Resolution

The radial resolution requirement of 30 meters is equivalent to a 0.2 microsecond pulse width. This value is decreased by a factor of two to account for adverse RF pulse relationship between two adjacent returns.

Equivalent pulse width $\tau_e = 0.1\ \mu s$

No angular resolution is specified.

False Alarm Probability

Allow one false alarm per second as a reasonable value for the system to handle without overtaxing the computer and radar on power and time spent on verification of false alarms.

$$\therefore P_N = \tau_e \text{ FAR} = 10^{-7} \text{s} \times 1/\text{s} = 10^{-7} \tag{1.37}$$

$$P_N = 10^{-7}$$

Search Pattern

Initially, select a beamwidth of 1° as suitable for the search and track modes. The eventual beamwidth selected will be based on obtaining acceptable values for search frame time, power-aperture product, aperture size and tracking performance. Several iterations may be required to arrive at a final design value. It may even be necessary to retain a 1° beamwidth for track and defocus the beam in search to obtain an acceptable frame time at a sacrifice in power.

The search sector will be scanned with interlaced instead of columned beams to decrease the depth of the holes between beam positions. (See diagram on following page.)

In addition, the pattern loss decreases as the beam stepping increment decreases but more beam positions and therefore more time is required to cover the search sector.

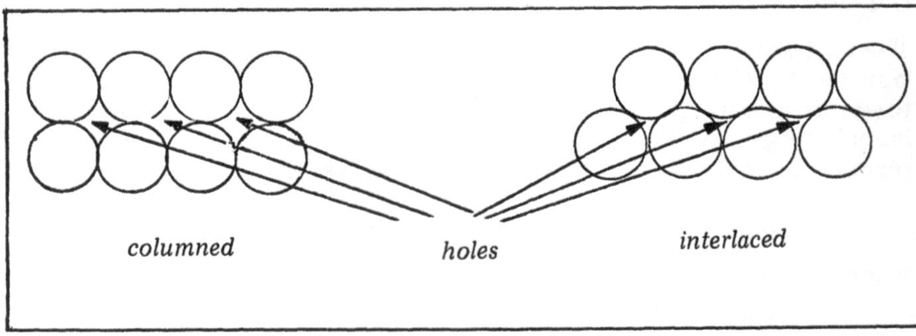

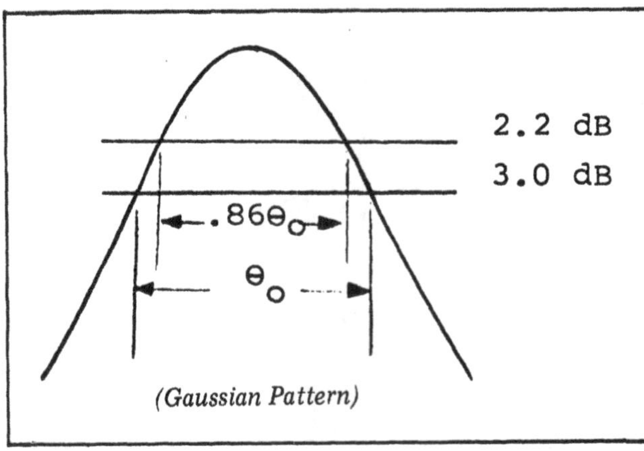

	Cross Over	Step Size	Pattern Loss
2.2 dB	3 dB	θ_o	3.2 dB
3.0 dB	2.2 dB	$.86\theta_o$	2.6 dB (typical)

The compromise selected in this design is to step the beam in full beamwidth (3 dB point) steps.

∴ Scan pattern: 17.45 millisine steps in azimuth
15.11 millisine steps in elevation

The configuration selected for array element spacing is an equilateral triangular grid structure to obtain conical coverage.

The array is oriented to face true north — direction of threat.

∴ $N_t = 0$

Determine tilt angle and maximum scan by iterating expression (7.9) for different tilt angles until the maximum scan angles at the lower and upper boundaries are equal.

Lower boundary $E_L = 0,\ A_{zL} = 45°$,

$\therefore \theta_L = \cos^{-1}(\cos 45° \cos T) = \cos^{-1}(.7071 \cos T)$

Upper boundary $E_U = 60°,\ A_{zU} = 45°$

$\therefore \theta_U = \cos^{-1}(\sin 60° \sin T + \cos 60° \cos 45° \cos T)$

$\theta_U = \cos^{-1}(.8660 \sin T + .3536 \cos T)$

T	22.0°	22.1°	22.2°	22.3°
θ_L	49.03°	49.07°	49.10°	49.14°
θ_U	49.29°	49.20°	49.11°	49.02°

$\therefore T = 22.2°$

$\xi_{max} = 49.1°$

The sine space expressions 7.1, 7.3, 7.5 and 7.6 required for determination of the search scan pattern for $N_t = 0°$, $T = 22.2°$ reduce to the following for each elevation beam position, n:

Beginning of scan $A_z = 0$

$$\alpha_{on} = \sin A_{on} = 0$$

$$\beta_{on} = \sin B_{on} = \sin(E_n - 22.2°) \qquad (14.1)$$

or

$$B_{on} = E_n - 22.2° \qquad (14.2)$$

End of scan $A_z = 45°$

$$\sin A_n = \alpha_n = .7071 \cos E_n \qquad (14.3)$$

$$\sin B_n = \beta_n = .9259 \sin E_n - .2672 \cos E_n \qquad (14.4)$$

$$\xi_n = \sin^{-1}\sqrt{\alpha_n^2 + \beta_n^2} \qquad (14.5)$$

Number of beam positions scanned ($\theta_o = 1°$)

$$N_{Bn} = \frac{2(\alpha_n - \alpha_{on})}{2\sin\frac{\theta_o}{2}} \approx \frac{2\alpha_n}{\theta_o} = \frac{2\alpha_n}{.01745} \qquad (14.6)$$

Procedure for determining search scan pattern:

1. Step the beam in elevation in .01511 sine space increments. Begin with the lower 3 dB point of the beam set on the horizon for elevation beam position # 1.

For each elevation beam position:

2. Find the maximum range from a Range-Height-Angle chart (Ref. 1); where the height is 50,000 feet and the elevation angle is determined from the lower 3 dB point of the beam. The maximum range used in the search pattern is the one obtained from the chart or 200 nm — whichever is the smallest.
3. Determine the sine space angles.
4. Determine the number of beam positions scanned.
5. Determine the maximum scan angle.
6. Terminate search in elevation at approximately 70 nm which is the minimum missile detection range.

\# 1

$$E_{L1} = 0°$$

$$R_{M1} = 200 \text{ nm (Range-Height-Angle Chart) (Ref. 1)}$$

$$B_{oL1} = E_{L1} - 22.2° = -22.2°$$

$$E_1 = \sin^{-1} (\sin B_{oL1} + \theta_o/2) + 22.2°$$

$$E_1 = \sin^{-1} (\sin -22.2° + .008725) + 22.2°$$

$$E_1 = -21.66° + 22.2° = 0.54°$$

$$\alpha_1 = .7071 \cos E_1 = .7071 \cos .54° = .7071$$

$$A_1 = \sin^{-1} (\alpha_1) = \sin^{-1} (.7071) = 45°$$

$$N_{B1} = \frac{2\alpha_1}{\theta_o} = \frac{2 \times .7071}{.01745} = 81$$

$$\beta_1 = .9259 \sin E_1 - .2672 \cos E_1$$

$$\beta_1 = .9259 \sin .54° - .2672 \cos .54° = -.2585$$

$$B_1 = \sin^{-1} (\beta_1) = \sin^{-1} (-.2585) = -14.98°$$

$$\xi_1 = \sin^{-1} \sqrt{\alpha_1^2 + \beta_1^2} = \sin^{-1} \sqrt{.7071^2 + .2585^2} = 48.84°$$

\#2

$$B_{oL2} = \sin^{-1} (\sin B_{oL1} + .01511)$$

$$B_{oL2} = \sin^{-1} (\sin - 22.2° + .01511) = -21.27°$$

$$E_{L2} = B_{oL2} + 22.2° = -21.27° + 22.2° = .93°$$

$$R_{M2} = 200 \text{ nm (Range-Height-Angle Chart)}$$

$E_2 = \sin^{-1}(\sin B_{oL2} + \theta_o/2) + 22.2°$

$E_2 = \sin^{-1}(\sin -21.27° + .008725) + 22.2° = 1.47°$

$a_2 = .7071 \cos E_2 - .7071 \cos 1.47° = .7069$

$A_2 = \sin^{-1}(a_2) = \sin^{-1} .7069 = 44.98°$

$$N_{B2} = \frac{2a_2}{\theta_o} = \frac{2 \times .7069}{.01745} = 81$$

$\beta_2 = .9259 \sin E_2 - .2672 \cos E_2$

$\beta_2 = .9259 \sin 1.47° - .2672 \cos 1.47° = -.2434$

$B_2 = \sin^{-1}(\beta_2) = \sin^{-1}(-.2434) = -14.08°$

$\xi_2 = \sin^{-1} \sqrt{a_1{}^2 + \beta_1{}^2} = \sin^{-1} \sqrt{.7069^2 + .2434^2} = 48.39°$

\# 3 $B_{oL3} = \sin^{-1}(\sin - 21.27° + .01511) = -20.34°$

$E_{L3} = -20.34° + 22.2° = 1.86°$

R_{M3} = 165 nm (Range-Height-Angle Chart)

$E_3 = \sin^{-1}(\sin - 20.34° + .008725) + 22.2° = 2.39°$

$a_3 = .7071 \cos 2.39° = .7065$

$A_3 = \sin^{-1} .7065 = 44.95°$

$$N_{B3} = \frac{2 \times .7065}{.01745} = 80.97$$

$\beta_3 = .9259 \sin 2.39° - .2672 \cos 2.39° = -.2284$

$B_3 = \sin^{-1}(-.2284) = -13.20°$

$\xi_3 = \sin^{-1} \sqrt{.7065^2 + .2284^2} = 47.95°$

\# 4 $B_{oL4} = \sin^{-1}(\sin - 20.34° + .01511) = -19.42°$

$E_{L4} = -19.42° + 22.2° = 2.78°$

R_{M4} = 134 nm

$E_4 = \sin^{-1}(\sin - 19.42° + .008725) + 22.2° = 3.31°$

$a_4 = .7071 \cos 3.31° = .7059$

$A_4 = \sin^{-1} .7059 = 44.90°$

$$N_{B4} = \frac{2 \times .7059}{.01745} = 80.91$$

$\beta_4 = .9259 \sin 3.31° - .2672 \cos 3.31° = -.2133$

$B_4 = \sin^{-1}(-.2133) = -12.32°$

$\xi_4 = \sin^{-1} \sqrt{.7059^2 + .2133^2} = 47.51°$

#5 $B_{oL5} = \sin^{-1} (\sin - 19.42° + .01511) = -18.50°$

$E_{L5} = -18.50° + 22.2° = 3.70°$

$R_{M5} = 110$ nm

$E_5 = \sin^{-1} (\sin - 18.5° + .008725) + 22.2° = 4.23°$

$a_5 = .7071 \cos 4.23° = .7052$

$A_5 = \sin^{-1} .7052 = 44.84°$

$$N_{B5} = \frac{2 \times .7052}{.01745} = 80.83$$

$\beta_5 = .9259 \sin 4.23° - .2672 \cos 4.23° = -.1982$

$B_5 = \sin^{-1} (-.1982) = -11.43°$

$\xi_5 = \sin^{-1} \sqrt{.7052^2 + .1982^2} = 47.10°$

6 $B_{oL6} = \sin^{-1} (\sin - 18.5° + .01511) = -17.59°$

$E_{L6} = -17.59° + 22.2° = 4.61°$

$R_{M6} = 95$ nm

$E_6 = \sin^{-1} (\sin - 17.59° + .008725) + 22.2° = 5.13°$

$a_6 = .7071 \cos 5.13° = .7043$

$A_6 = \sin^{-1} .7043 = 44.77°$

$$N_{B6} = \frac{2 \times .7043}{.01745} = 80.72$$

$\beta_6 = .9259 \sin 5.13° - .2672 \cos 5.13° = -.1833$

$B_6 = \sin^{-1} (-.1833) = -10.56°$

$\xi_6 = \sin^{-1} \sqrt{.7043^2 + .1833^2} = 46.70°$

7 $B_{oL7} = \sin^{-1} (\sin - 17.59° + .01511) = -16.68°$

$E_{L7} = -16.68° + 22.2° = 5.52°$

$R_{M7} = 82$ nm

$E_7 = \sin^{-1} (\sin -16.68° + .008725) + 22.2° = 6.04°$

$a_7 = .7071 \cos 6.04° = .7032$

$A_7 = \sin^{-1} .7032 = 44.68°$

$$N_{B7} = \frac{2 \times .7032}{.01745} = 80.60$$

$\beta_7 = .9259 \sin 6.04° - .2672 \cos 6.04° = -.1683$

$B_7 = \sin^{-1}(-.1683) = -9.69°$

$\xi_7 = \sin^{-1}\sqrt{.7032^2 + .1683^2} = 46.31°$

\# 8 $B_{oL8} = \sin^{-1}(\sin - 16.68° + .01511) = -15.78°$

$E_{L8} = -15.78 + 22.2° = 6.42°$

$R_{M8} = 72$ nm ∴ Last elevation beam position

$E_8 = \sin^{-1}(\sin - 15.78° + .008725) + 22.2° = 6.94°$

$a_8 = .7071 \cos 6.94° = .7019$

$A_8 = \sin^{-1} .7019 = 44.58°$

$$N_{B8} = \frac{2 \times .7019}{.01745} = 80.45$$

$\beta_8 = .9259 \sin 6.94° - .2672 \cos 6.94° = -.1534$

$B_8 = \sin^{-1}(-.1534) = -8.82°$

$\xi_8 = \sin^{-1}\sqrt{.7019^2 + .1534^2} = 45.93°$

The sine space coverage pattern is plotted in Fig. 14.1 and summarized in Table 14.1, Parameter Management Chart. The Chart contains a tabulation of all the parameters required to determine the time budget as well as the derivation of the time budget.

Probability of Detection

Aircraft — SW I

Design for $P_c = 99\%$ $P_N = 10^{-7}$

Frame Time = 2.42s (refer to Table 14.1)

V = MACH 2 = 2 x .184 nm/s = .368 nm/s

Radial range increment per frame time = .368 x 2.42 = .89 nm

$$P_c = 1 - \prod_{i=1}^{M} (1 - P_{D_i}) \qquad (4.44)$$

Let $(S/N)_M = 10.5$ dB at maximum range of 200 nmi

(This calculation may require several reiterations)

∴ $P_{D_1} = .275$ (from probability charts for SW I)

$P_N = 10^{-7}$ (Refs. 2 and 3)

The total number of such beams is 8 E x 81 A_z = 648

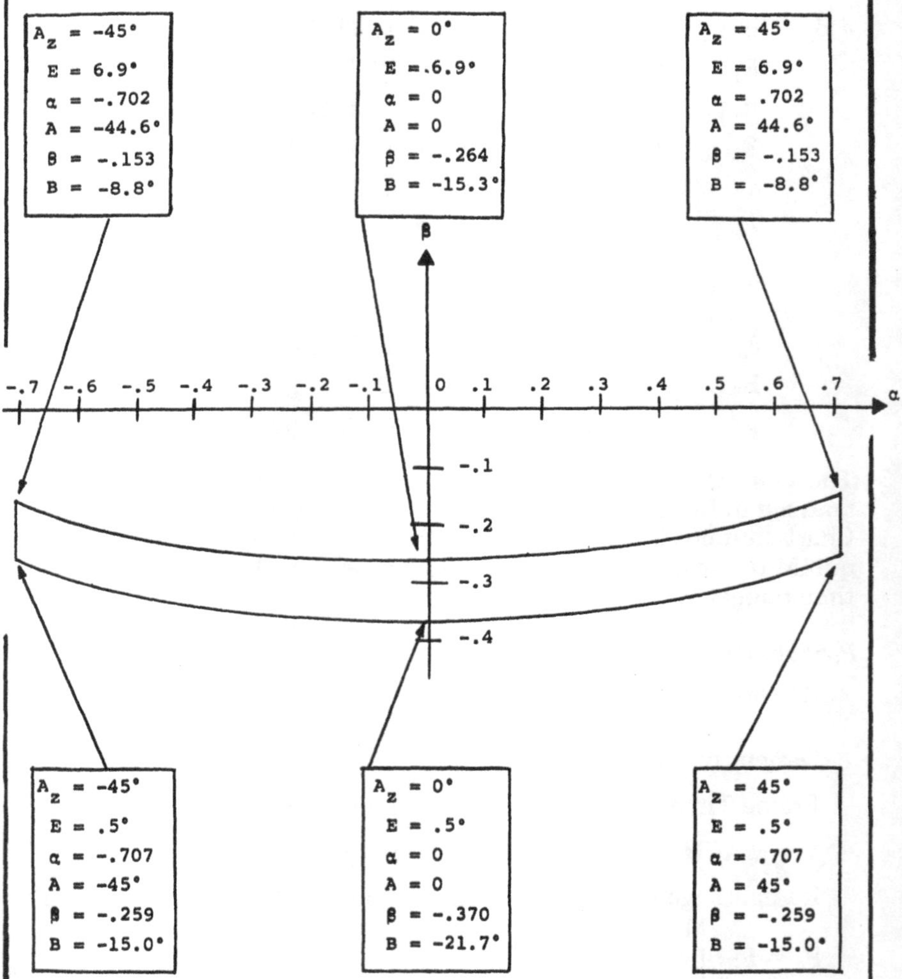

Figure 14-1

Search Coverage Pattern

At other ranges $(S/N)_{i_{dB}} = (S/N)_{M_{dB}} + \left(\frac{R_M}{R_i}\right)^4_{dB} + \Delta L^*_{ATi_{dB}}$ (4.46)

i	R_i nm	$(S/N)_i$ dB	P_{D_i}	$1-P_{D_i}$	$\pi(1-P_{D_i})$	P_{c_i}	
1.	200.00	10.50	.275	.725	.725	.275	
2.	199.11	10.58	.280	.720	.522	.478	
3.	198.22	10.66	.290	.710	.371	.629	
4.	197.33	10.73	.295	.705	.262	.738	
5.	196.44	10.81	.300	.700	.183	.817	
6.	195.55	10.89	.305	.695	.127	.873	
7.	194.66	10.97	.310	.690	.088	.912	
8.	193.77	11.05	.312	.688	.061	.939	
9.	192.88	11.13	.320	.680	.041	.959	
10.	191.99	11.21	.330	.670	.021	.973	
11.	191.10	11.29	.335	.665	.018	.982	
12.	190.21	11.37	.340	.660	.012	.988	
13.	189.32	11.45	.345	.655	.008	.992	OK

* ΔL_{AT_i} = Decrease in atmospheric loss between Ranges R_i and R_M (not included in these calculations).

Let $(S/N)_M$ = 10.0 dB and reiterate calculation.

n i	R_i nm	$(S/N)_i$ dB	P_{D_i}	$1-P_{D_i}$	$\pi(1-P_{D_i})$	P_{c_i}
1.	200.00	10.00	.240	.760	.760	.240
2.	199.11	10.08	.245	.755	.574	.426
3.	198.22	10.16	.250	.750	.430	.570
4.	197.33	10.23	.255	.745	.320	.680
5.	196.44	10.31	.260	.740	.237	.763
6.	195.55	10.39	.270	.730	.173	.827
7.	194.66	10.47	.275	.725	.125	.875
8.	193.77	10.55	.278	.722	.090	.910

9.	192.88	10.63	.285	.715	.058	.942	
10.	191.99	10.71	.290	.710	.041	.959	
11.	191.10	10.79	.295	.705	.021	.979	
12.	190.20	10.87	.300	.700	.020	.980	
13.	189.32	10.95	.310	.690	.014	.986	
14.	188.43	11.04	.312	.688	.010	.990	OK

$\therefore R_{A/C} = 188$ nm (> 180 nm OK)

The above reiteration indicates another 0.5 dB has been saved in S/N at the expense of 0.89 nm decrease in detection range. Practical experience has indicated that a single-hit signal-to-noise ratio of 10.0 to 11.0 dB at the desired detection range is a reasonable minimum for ESRS operation.

Missile — SW II

$V_{closing}$ = MACH 3 = 3 x .184 = .552 nm/s

Radial range increment per frame time = .552 x 2.42 = 1.34 nm

@ 100 nm 3 dB range improvement $\left(\frac{200}{100}\right) \times 4 = 12.0$ dB

Cross section loss $\sigma_{A/C}/\sigma_{MSL} = \frac{1.0}{.05} = 20 = 13.0$ dB

$\therefore$ Net loss = 1.0 dB

(Conservative because decrease in atmospheric loss has not been included)

n i	R_i nm	$(S/N)_i$ dB	P_{D_i}	$1-P_{D_i}$	$\pi(1-P_{D_i})$	P_{c_i}
1.	100.00	9.00	.150	.850	.850	.150
2.	98.66	9.23	.165	.835	.710	.290
3.	97.32	9.47	.180	.820	.582	.418
4.	95.98	9.71	.205	.795	.463	.537
5.	94.64	9.96	.225	.775	.359	.641
6.	93.30	10.20	.245	.755	.271	.729
7.	91.96	10.46	.270	.730	.198	.802

8.	90.62	10.71	.285	.715	.142	.858	
9.	89.28	10.97	.315	.685	.097	.903	
10.	87.94	11.23	.340	.660	.064	.936	
11.	86.60	11.50	.370	.630	.040	.960	
12.	85.26	11.77	.400	.600	.024	.976	
13.	83.92	12.05	.430	.570	.014	.986	
14.	82.58	12.33	.460	.540	.008	.992	OK

$\therefore R_{MSL} = 82$ nm (> 70 nm OK)

For the target not to move from one beam to the next beam:

$$\left(\text{Frame Time} \leqslant \frac{R_{min}\theta_o}{V_{max}} = \frac{82 \times .01745}{.552} = 2.59 \text{ s}\right)$$

($\therefore$ 2.42 s $\leqslant$ 2.59 s OK)

Energy Requirements

The system parameters are determined for various ESRS and a cost-effectiveness study is then performed to select the most suitable design for this application. The method will be demonstrated for an optically-fed thinned array.

$$\mathcal{E} = \frac{S/N\ R^4\ L(NF)_0}{\nu\ G^2\ \sigma\ \lambda^2} \tag{1.27}$$

S/N = 10.0 dB } from A/C

R^4 = 200^4 nm = 92.04 dB nm } P_c = 99% calculations

L = total losses not included in system noise temperature

L_{atm} = 3.3 dB (R.T. @ 0.5°E, 3 GHz, 200 nmi) (Ref. 4)

Antenna (Refer to Table 11.1)	Non-Ohmic	Ohmic
Radome		0.5 dB
Collector Illumination taper	0.2 dB	
Spillover	1.2	
Feed		0.2
3 bit Phase Shifter Insertion loss		1.2
3 bit Quantization	0.2	

Element mismatch (front & rear)			0.3
Element insertion loss			0.2
Phase tolerance -20° RMS	0.4		
Manufacturing tolerance	0.3	2.3 dB	2.4 dB
Duplexer	0.5 dB		
Waveguide	0.2		
Isolator	0.3		
Harmonic filter	0.1		
Mismatched filter	1.5	2.6 dB	
Beam pattern (3 dB steps)			3.2 dB
Beam Scanning $\sec^3 49° = 3.54$ =			5.5 dB
L = 3.3 + 4.7 + 2.3 + 2.6 + 3.2 + 5.5 =			21.6 dB

System Noise Temperature

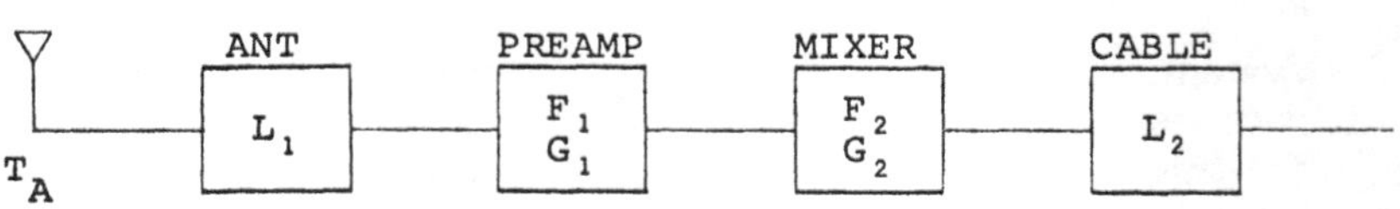

$$T_S = T_A + T_O \left[(L_1 - 1) + L_1 (F_1 - 1) + L_1 \frac{(F_2 - 1)}{G_1} + \frac{L_1 (L_2 - 1)}{G_1 G_2} \right]$$

T_A = 90°K (0.5°E, 3 GHz) (Ref. 4)

T_O = 290°K

L_1 = L_{ANT} 2.4 dB (ohmic)

Duplexer 0.5

TR Switch 1.0

3.9 dB = 2.45

F_1 = 2.0 db = 1.58 (uncooled parametric amplifier)

G_1 = 30 dB = 1000

F_2 = 8.0 dB = 6.31 mixer

G_2 = -6.0 dB = .25

L_2 = 3.0 dB = 2.0 cable

$$\therefore\ T_s = 90 + 290\left[\underbrace{2.45\text{-}1}_{1.45} + \underbrace{2.45 \times (1.58\text{-}1)}_{1.42} + \underbrace{\frac{2.45(6.31\text{-}1)}{1000}}_{.0130} + \underbrace{\frac{2.45(2.0\text{-}1)}{1000 \times .25}}_{.0098}\right]$$

$$T_s = 90 + 290 \times 2.893 = 90 + 839 = 929^\circ K$$

$$(NF)_o = \frac{T_s}{T_o} = \frac{929}{290} = 3.203 = 5.06 \text{ dB}$$

$$\nu = 0.28 \text{ dB (expression 1.31)}$$

$$G = \frac{4\pi A_o}{\lambda^2}$$

$$D = \frac{71.3\lambda}{BW}$$ (Taylor taper 35 dB max sidelobe $\bar{n} = 5$) (Circular array) (Ref. 7, Page 256)

$$\lambda = \frac{c}{f_o} = \frac{3 \times 10^{10} \text{ cm/s}}{3 \times 10^9 \text{ Hz}} = 10 \text{ cm}$$

$$D = \frac{71.3 \times 10}{1^\circ} = 713 \text{ cm} = 7.13 \text{ m} = 23.4'$$

$$A_o = .7854D^2 = .7854 \times 713^2 = 3.99 \times 10^5 \text{ cm}^2$$

Hexagonal fit on circular array

Maximum scan angle $\xi_{max} = 49^\circ$

Allowing 10° margin, design for scan of 59°

$$d_e = \frac{1}{1 + \sin\xi} = \frac{1}{1 + \sin 59^\circ} = \frac{\lambda}{1.857} = .538\lambda = 5.38 \text{ cm}$$

For element spacing on an equilateral triangle grid (expression 7.60)

$$a = 1.1547 d_e^2 = 1.1547 \times 5.38^2 = 33.5 \text{ cm}^2$$

$$\text{Number of elements} = \frac{A_o}{a} = \frac{3.99 \times 10^5}{33.5} = 11910$$

Use 50% thinning

$\therefore$ No. of active elements = 5955

No. of dummy elements = 5955

$$G = \frac{4\pi \times 3.99 \times 10^5}{100} = 5.01 \times 10^4 = 47.00 \text{ dB}$$

$$\text{-50\% thinning} = \underline{-3.00}$$

$$44.00 \text{ dB}$$

G^2 = 88.00 dB

σ = 1 m^2 = 0 dB m^2

λ = 100 cm^2 = 20 dB cm^2

$$\mathcal{E} = \frac{S/N\ R^4\ L(NF)_o}{\nu\ G^2\ \sigma\ \lambda^2}$$

S/N	= 10.00	ν	= 0.28
R^4	= 92.04	G^2	= 88.00
L	= 21.60	σ	= 0.00
$(NF)_o$	= 5.06	λ^2	= 20.00
	128.70		108.28
	-108.28		

$\mathcal{E}$ = 20.42 dB = 110.2 Joules

R_{min} = 10 nm = 123.6 μs

$\therefore$ T_d = 110 μs (< 123.6 μs) (pulse duration)

$$\therefore\ P_p = \frac{\mathcal{E}}{T_d} = \frac{110.2}{110 \times 10^{-6}} = 1 \text{ Mw}$$

14.3 Track

Waveform

Let the waveform be the same for search and track as a first iteration. It makes for a simpler system.

T_d = 110 μs/0.1 μs (1100:1 linear fm pulse compression)

Signal-to-noise ratio

$$S/N = 11.0 \text{ dB @ } R = 188 \text{ nm}$$

The signal-to-noise ratio is approximately the same for search and track. The range gate loss is offset by the improvement in beam shape loss. In track the target is closer to the peak of the beam position.

Accuracy — Angle (Refer to Table 9.1)

1. Antenna pointing (9.1)

$$\sigma_b = \frac{\xi \theta_o}{2\sqrt{N_a}}$$

ξ = 20° (rss of phase errors)
(Array designer evaluation)

θ_o = 1° (3 dB beamwidth)

N_a = 5955 (number of active elements)

$$\therefore \sigma_b = \frac{20 \times .01745 \times .01745}{2\sqrt{5955}} = 3.95 \times 10^{-5} = .04 \text{ ms}$$

2. Data Quantization

$$\sigma_{QA} = \frac{2\pi}{\sqrt{12}\ 2^Q} \tag{9.2}$$

Q = 16 (Computer X word length)

$$\sigma_{QA} = \frac{2\pi}{\sqrt{12}\ 2^{16}} = 2.77 \times 10^{-5} = .03 \text{ mils}$$

3. A/D Conversion = .05 ms (A/D circuit designer evaluation)

4. Thermal

$$\sigma_{TA} = \frac{\theta_o}{K_M \sqrt{2\ S/N}} \tag{9.3}$$

K_M = 1.8 (array designer evaluation)

S/N = 11.0 dB = 12.59

$$\sigma_{TA} = \frac{.01745}{1.8\sqrt{12.59 \times 2}} = 1.93 \times 10^{-3} = 1.93 \text{ ms}$$

5. Instrumentation

4 horn monopulse feed with 3 channel single-hit normalization (limiting receiver)

	Errors Amplitude	Phase
Precomparator	0.5 dB	5°
Post comparator	0.5 dB	10°
Post post comparator (limiter)	—	2°

$$\sigma_I = 1 - G_1 \left[\frac{G_2 \cos(a_1 + a_2 + a_3) + \sin(a_2 + a_3)}{G_2 \cos(a_2 + a_3) - \sin a_3} \right] S_o \qquad (9.6)$$

$a_1 = 5° \qquad a_2 = 10° \qquad a_3 = 2°$

$G_1 = G_2 = 0.5\text{ dB} = 1.06$

$\theta_o = .01745 \qquad K_M = 1.8$

$G = 44.0 - 4.7 = 39.3\text{ dB} = 8511$

$$S_o = \frac{\theta_o}{K_M \sqrt{G}} = \frac{.01745}{1.8\sqrt{8511}} = 1.05 \times 10^{-4}$$

$$\sigma_I = [\,.06^2 + .0834^2 + .0724^2\,]^{1/2}\, S_o$$

$$\sigma_I = .1257 \times 1.05 \times 10^{-4} = 1.32 \times 10^{-5} = .0132\text{ ms}$$

6. Dynamic Lag

Design for a type II sample data servo loop to the computer for all 3 sine space coordinates for both aircraft and missile targets.

Use one sample per second data rate for aircraft targets and two samples per second data rate for missile targets.

Determine if the smoothing constants of a critically damped system are satisfactory and then compute the dynamic lag error.

$$\epsilon_p = \frac{a_m t^2}{\beta_{CD}} \qquad (8.1)$$

where:

ϵ_D = dynamic lag error

$\beta_{CD} < 4$ (for stability)

$\alpha_{CD} = 2\sqrt{\beta_{CD}} - \beta_{CD}$

$\epsilon_D = \epsilon_p (1\text{-}\alpha_{CD})$ (9.7)

ϵ_p = error in predicted position

a_m = maximum acceleration of tracked target

t = sample period

β_{CD} = velocity smoothing constant

α_{CD} = position smoothing constant

Aircraft

Maximum angular acceleration occurs at minimum range

$$a_m = 3 \text{ gs} = 3 \times 32.2 = 96.6 \text{ fps}$$

$$V_m = M2 = 2 \times .184 = .368 \text{ nm/s}$$

$$R_{min} = 10 \text{ nm}$$

$$\ddot{\phi}_1 = .65\left(\frac{V_m}{R_{min}}\right)^2 = \frac{.65 \times .368^2}{10^2} \times 10^3 = .88 \text{ mils/s}^2$$

$$\ddot{\phi}_2 R_{min} = a_m$$

$$\therefore \quad \ddot{\phi}_2 = \frac{96.6}{10 \times 6080} \times 10^3 = 1.59 \text{ mils/s}^2$$

$$\epsilon_p = \frac{17.45}{2} \text{ mils} = 8.73 \text{ mils } (\tfrac{1}{2} \text{ beamwidth})$$

$$t = 1.0 \text{ second}$$

$$\therefore \alpha_m = 1.59 \text{ mils/s}^2$$

$$\beta_{CD} = \frac{aT^2}{\epsilon_p} = \frac{1.59 \times 1^2}{8.73} = .182 \ (< 4 \text{ OK})$$

$$\alpha_{CD} = \underbrace{2\sqrt{.182}}_{.853} - .182 = .671$$

$$\epsilon_D = 8.73\,(1\text{-}.671) = 2.87 \text{ ms}$$

Missile

$$a_m = 10 \text{ gs} = 10 \times 32.2 = 322 \text{ fps}$$

$$V_m = M3 \text{ (relative to ground)} = 3 \times .184 = .552 \text{ nm/s}$$

$$R_{min} = 10 \text{ nm}$$

$$\ddot{\phi}_1 = \frac{.65 \times .552^2}{100} \times 10^3 = 1.98 \text{ mils/s}^2$$

$$\ddot{\phi}_2 = \frac{322}{10 \times 6080} \times 10^3 = 5.30 \text{ mils/s}^2$$

$$\epsilon_p = 8.73 \text{ mils}$$

$$t = 0.5 \text{ seconds}$$

$$\therefore \alpha_m = 5.30 \text{ mils/s}^2$$

$$\beta_{CD} = \frac{5.30 \times 0.5^2}{8.73} = 0.152 \quad (< 4 \text{ OK})$$

$$\alpha_{CD} = \underbrace{2\sqrt{.152}}_{.780} - 0.152 = 0.628$$

$$\epsilon_D = 8.73 \,(1\text{-}.628) = 3.25 \text{ ms}$$

7. Scan

$$\sigma_s \approx \frac{\theta_o}{40} = \frac{.01745}{40} = 4.36 \times 10^{-4} = 0.4 \text{ ms}$$

(array designer evaluation)

8. Multipath

$$\sigma_{MA} = \frac{\theta_o \rho}{\sqrt{8\, G_{SL}}} \tag{9.8}$$

$$G_{SL} = \frac{5}{N\, 2^{2P}} \tag{14.7}$$

(Ref. 5)

$$G_{SL} = \frac{5}{5955 \times 2^6} = 1.31 \times 10^{-5} \quad (-49 \text{ dB})$$

(rms sidelobe level)

OR

G_{SL} = 28 dB = 631 (maximum sidelobe level) (array designer evaluation)

use 631

ρ = 0.5 (smooth surface reflectivity)

$$\sigma_{M_A} = \frac{.01745 \times 0.5}{\sqrt{8 \times 631}} = 1.23 \times 10^{-4} = 0.1 \text{ ms}$$

9. Propagation and Calibration

Aircraft

$\sigma_{PG_A} \approx .25$ mils (system designer evaluation)

Missile

$\sigma_{PG_A} \approx .13$ mils (See Table 14.2 on following page.)

Accuracy — Range (Refer to Table 9.2)

1. Data quantization

$$\sigma_{QR} = \frac{LSB}{\sqrt{12}} \tag{9.18}$$

Let LSB = 1 m

$$\sigma_{QR} = \frac{1}{\sqrt{12}} = 0.29 \text{ m}$$

2. Pulse jitter

$$\sigma_{PR} = \frac{c\Delta t_s}{2\sqrt{12}} \tag{9.19}$$

$\Delta t_s =$ 10 ns (transmitter designer evaluation)

$$\sigma_{PR} = \frac{3 \times 10^8 \times 10^{-8}}{2\sqrt{12}} = 0.43 \text{ m}$$

3. Range clock

$$\sigma_c = \frac{c}{2\sqrt{12}\, f_o} \tag{9.20}$$

Use 80 MHz crystal oscillator

$$\therefore \sigma_c = \frac{3 \times 10^8}{2\sqrt{12} \times 8 \times 10^7} = 0.54 \text{ m}$$

4. FM Waveform

Table 14.2
Angle Errors

(one sigma values)

		Aircraft (ms)	Missile (ms)
1.	Antenna pointing	0.04	0.04
2.	Data quantization	0.03	0.03
3.	A/D conversion	0.05	0.05
4.	Thermal	1.93	1.93
5.	Instrumentation	0.01	0.01
6.	Dynamic lag	2.87	3.25
7.	Scan	0.40	0.40
8.	Multipath	0.01	0.01
9.	Propagation & Calibration	0.25	0.13
	Total rss	3.49	3.80

$$\sigma_{FM} = \frac{T_o V_M f_o}{\sqrt{12}\ \Delta f} \qquad (9.21)$$

$T_o = 110 \times 10^{-6}$

$V_M = 3 \times 340.5 \text{ m/s} = 1021.5 \text{ m/s}$

$f_o = 3 \times 10^9 \text{ Hz}$

$\Delta f = 10^7 \text{ Hz}$

$$\sigma_{FM} = \frac{110 \times 10^{-6} \times 1021.5 \times 3 \times 10^9}{\sqrt{12} \times 10^7} = 9.73 \text{ m}$$

For aircraft:

$\dot{R}$ correction reduces this value to 2.5 m

For missile:

$V_M = 4 \times 340.5 = 1362 \text{ m/s}$

$$\sigma_{FM} = \frac{1362}{1021.5} \times 9.73 = 12.97 \text{ m}$$

$\dot{R}$ correction reduces this value to 3.0 m

5. A/D conversion = .15 m (circuit designer evaluation)

6. Thermal

$$\sigma_{TR} = \frac{c\tau_e}{2\sqrt{2\ S/N}} \qquad (9.22)$$

$\tau_e = 0.1\ \mu s$ (compressed pulse width)

$S/N = 11.0 \text{ dB} = 12.59$

$$\sigma_{TR} = \frac{3 \times 10^8 \times 10^{-7}}{2\sqrt{2 \times 12.59}} = 2.99 \text{ m}$$

7. Dynamic Lag

Aircraft

$t = 1.0$ seconds

$\epsilon_p = 7.5\text{m}$ (½ compressed pulse width)

$$a_m = 3 \text{ gs} = \frac{3 \times 32.2}{3.281} = 29.44 \text{ m/s}^2$$

$$\beta_{CD} = \frac{29.44 \times 1^2}{7.5} = 3.9256 \ (< 4 \ \text{OK})$$

$$\alpha_{CD} = \underbrace{2\sqrt{3.925}}_{3.9626} - 3.9256 = .0370$$

$$\epsilon_D = 7.5\ (1\text{-}.0370) = 7.22 \text{ m}$$

Missile

$$t = 0.5 \text{ seconds}$$

$$\epsilon_p = 7.5 \text{ m}$$

$$a_m = 10 \text{ gs} = \frac{10 \times 32.2}{3.281} = 98.14 \text{ m/s}^2$$

$$\beta_{CD} = \frac{98.14 \times 5^2}{7.5} = 3.271$$

$$\alpha_{CD} = \underbrace{2\sqrt{3.271}}_{3.617} - 3.271 = .3464$$

$$\hat{\epsilon}_D = 7.5\ (1\text{-}.346) = 4.90 \text{ m}$$

8. Multipath

$$\sigma_{M_R} = \frac{c\rho\tau_e}{2\sqrt{8\ G_{SL}}} \tag{9.24}$$

$$\sigma_{M_R} = \frac{3 \times 10^8 \times 0.5 \times 10^{-7}}{2\sqrt{8 \times 631}} = .11 \text{ m}$$

9. Propagation and Calibration

Aircraft

$$\sigma_{PG_R} \approx 2.0 \text{ m} \quad \text{(system designer evaluation)}$$

Missile

$$\sigma_{PG_R} \approx 1.0 \text{ m}$$

Table 14.3
Range Errors

(one sigma values)

		Aircraft (m)	Missile (m)
1.	Data quantization (LSB=1 m)	0.29	0.29
2.	Pulse jitter	0.43	0.43
3.	Range clock (80 MHz)	0.54	0.54
4.	FM waveform — velocity uncertainty	2.50	3.00
5.	A/D conversion	0.15	0.15
6.	Thermal	2.99	2.99
7.	Dynamic lag	7.22	4.90
8.	Multipath	0.11	0.11
9.	Propagation & Calibration	2.00	1.00
	Total rss	8.48	6.60

14.4 MTI — Moving Target Indication

The MTI processor is designed to provide an adequate improvement factor on distributed ground clutter for the full MTI coverage range for all the azimuth beams at the lowest two elevation beam positions.

The maximum range of clutter return is 100 Km

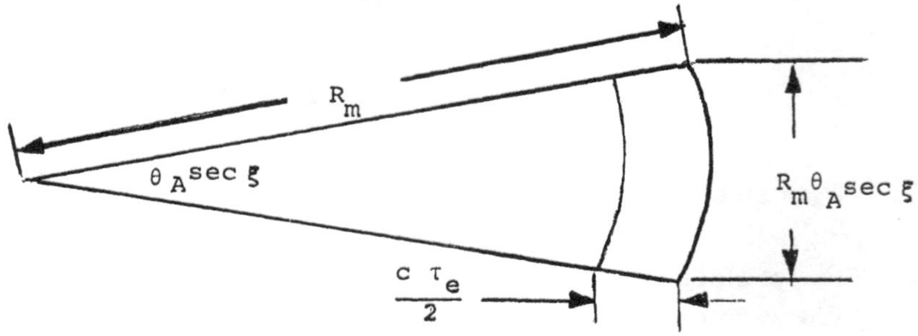

$$\text{Maximum reflected clutter area } \sigma_c = \sigma_o \frac{c\tau_e}{2} \frac{R_m \theta_A \sec\xi}{L_{BS}} \qquad (6.16)$$

where:

$$\sigma_o = \frac{.00032}{\lambda_o} \quad \text{(section 6.3)}$$

$$\lambda_o = 0.1 \text{ m} \quad (f_o = 3 \text{ GHz})$$

$$\sigma_o = \frac{.00032}{0.1} = 3.2 \times 10^{-3} \quad (-25 \text{ dB})$$

$$\sigma_c = \frac{3.2 \times 10^{-3} \times 3 \times 10^{8} \times 10^{-7}}{2} \times \frac{10^{5} \times .01745 \sec 49^\circ}{1.33}$$

$$\sigma_c = 95.99 \text{ m}^2 = 19.82 \text{ dB m}^2$$

Required Improvement Factor

$$I_{R_{dB}} = (S/N)_{o_{dB}} + \frac{\sigma_c}{\sigma_{t_{dB}}} + 6 \text{ dB} \qquad (6.1)$$

$(S/N)_o = 9.0$ dB $\quad \sigma_t = .05 \text{ m}^2$

$$\therefore \; I_{R_{dB}} = 9.0 + \log \frac{95.99}{.05} + 6 \qquad \text{(missile-worse case)}$$

$$I_{R_{dB}} = 9.0 + 32.83 + 6 = 47.83 \text{ dB}$$

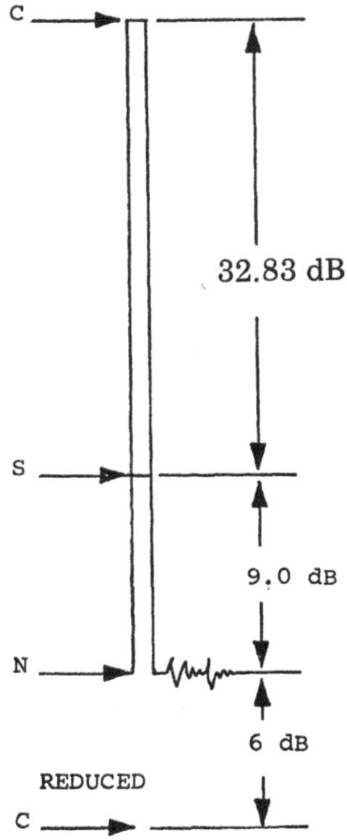

Available Improvement Factor

$$I_{2P}^{-1} = \frac{79\sigma_v^2 t_o^2}{\lambda_o^2}\left[1 - \frac{39.5\sigma_v^2 t_o^2}{\lambda_o^2}\right] \qquad (6.7)$$

$$\sigma_v = \frac{c}{\sqrt{8a}} \text{ m/s} \quad \text{(velocity spread)}$$

$$a = 2.3 \times 10^{17} \quad \text{(heavily wooded hills)}$$

(Ref. 6)

$$\sigma_v = \frac{3 \times 10^8}{\sqrt{8 \times 2.3 \times 10^{17}}} = 0.22 \text{ m/s}$$

$$t_o = 667\,\mu s \quad (100 \text{ K m})$$

$$+125\,\mu s \quad \text{(retrace time)}$$

$$t_o = 792\,\mu s = 7.92 \times 10^{-4} \text{ s}$$

$$I_{2P}^{-1} = \underbrace{\frac{79 \times .22^2 \times (7.92 \times 10^{-4})^2}{0.1^2}}_{2.4 \times 10^{-4}} \left[1 - \underbrace{\frac{2.4 \times 10^{-4}}{2}}_{\text{negligible}}\right]$$

$$I_{2P} = \frac{1}{2.4 \times 10^{-4}} = 4.17 \times 10^3 = 36.2 \text{ dB} \; (> 47.8 \text{ dB inadequate})$$

Double Canceller

$$I_{3P}^{-1} = \frac{1.25 \times 10^4 \sigma_v^4 t_o^4}{\lambda_o^4} + \frac{105\sigma_v^2 \delta^2}{\lambda_o^2} \qquad (6.8)$$

Use 10% stagger $\therefore \; \Delta = 0.1$

$$\Delta = 1 - \frac{t_o - \delta}{t_o + \delta} \qquad (14.8)$$

$$\delta = \frac{\Delta t_o}{2 - \Delta}$$

$$\delta = \frac{0.1\, t_o}{2 - 0.1} = .0526 t_o$$

$$I_{3P}^{-1} = ½(2.4 \times 10^{-4})^2 + \frac{105 \times .22^2 \times .0526^2 \times (7.92 \times 10^{-4})^2}{0.1^2}$$

$$I_{3P}^{-1} = 2.88 \times 10^{-8} + 88.20 \times 10^{-8} = 91.08 \times 10^{-8}$$

$$I_{3P} = \frac{1}{91.08 \times 10^{-8}} = 1.098 \times 10^6 = 60.4 \text{ dB } (> 47.8 \text{ dB OK})$$

MTI design selection is double canceller (10% stagger) with 5 pulses per beam position (3 independent residues)

Blind Speeds

With 10% staggering, blind speeds occur at 10 times the pulse repetition frequency.

$$\therefore f_{BS} = 10 \times f_r = \frac{10}{t_o} = \frac{10}{7.92 \times 10^{-4}} = 1.263 \times 10^4 \text{ Hz}$$

$$V_{BS} = \frac{f_{BS}\,\lambda_o}{2} = \frac{1.26 \times 10^4 \times 10^{-1}\text{ m}}{2} = 631 \text{ m/s} = \text{M } 1.85$$

Rejection Velocities

Assuming the threshold level is 30 dB below the $\sin^2 \phi$ function of a 3 pulse canceller combed filter response:

$$\sin^2 \phi = 30 \text{ dB}$$

$$\sin \phi = 15 \text{ dB (voltage)} = .1778$$

$$\phi = 10.24°$$

$$\frac{10.24}{180} \times 63.1 \text{ m/s} = 3.59 \text{ m/s} = \text{M } .0105 = 7 \text{ knots}$$

$\therefore$ Reject 0-7 Knots

14.5 Parameter Management

1. Time

1) Search

The maximum range as bounded by 50000 feet altitude is a function of the earth's curvature, atmospheric refraction and elevation beam position.

The interpulse period is decreased as the beam is stepped in elevation. The time interval corresponds to the maximum range for each elevation beam position. The maximum range is determined from charts using the lower 3 dB angle of the elevation beam. (Ref. 1)

2) Track

The track sampling rate can be decreased as the target dynamics decrease. The error signal in range (and angle) can be monitored to control the sampling rate on each tracked target.

2. Energy

1) Search

The energy is decreased in 3 dB increments (The reason for use of binary increments will become evident when waveform management is discussed.) as the maximum range decreases with elevation beam position as determined by the following formula:

$$\epsilon R_{n_{dB}} = \left(\frac{R_i}{R_n}\right)^4_{dB} + \Delta ATD_{dB} \qquad (14.9)$$

where

ϵR_n = energy reduction at n^{th} elevation beam position

R_i = initial range where the process begins

R_n = maximum range at n^{th} elevation beam position

ΔATD = decrease in atmospheric loss between R_i and R_n (Ref. 4)

Energy management is usually determined relative to the lowest elevation beam position; but in the Air Defense ESRS full energy is required from 200 nm to 82 nm to obtain the required cumulative probability of detection for the missile. Consequently the energy can only be reduced in elevation beam position # 8 (refer to Table 14.1).

8

$$\epsilon R_s = \left(\frac{82}{72}\right)^4_{dB} + (3.3 - 0.8)_{dB} \text{ (Ref. 4)} \qquad (E = 6.4°)$$

$$\epsilon R_s = 2.26 + 2.5 = 4.76 \text{ dB}$$

Use 3 dB = 0.5 (binary increment)

$$\therefore \quad \epsilon R_s = 110 \times 0.5 = 55 \text{ joules}$$

(Without the missile detection requirement much more energy can be saved by using energy management in elevation beam positions # 3 through # 8 with R_i equal to 200 nm.)

2) Track
The energy is decreased in binary increments as the range, R_T, of each tracked target decreases as determined by the formula:

$$\epsilon R_T = \left(\frac{200}{R_T}\right)^4 \qquad \text{(aircraft targets only)}$$

3) MTI
The energy required for MTI can be reduced because the maximum MTI range is 54 nm.

$$\therefore \epsilon R_{MTI} = \left(\frac{82}{54}\right)^4_{dB} + (3.3 - 1.5)_{dB} \text{ (Ref. 4)} \qquad (E = 0.5°)$$

$$= 7.26 + 1.8 = 9.06 \text{ dB}$$

Use 9 dB = 0.125

$$\therefore \epsilon_{MTI} = 110 \times 0.125 = 13.75 \text{ joules}$$

4) Beam scanning
The beam scanning loss decreases as the scan angle off broadside decreases. In the region where $\sec^3 \xi \leqslant 2.5$ db the energy can be reduced by a binary increment. This saving applies to search, track and MTI operational modes.

$$\sec^3 \xi_{max_{dB}} - \sec^3 \xi_{0_{dB}} = 3.0 \text{ dB}$$

$$\underbrace{\sec^3 49°}_{5.5 \text{ dB}} - \sec^3 \xi_{0_{dB}} = 3.0 \text{ dB}$$

$$\sec^3 \xi_0 = 2.5 \text{ dB}$$

OR

$$\sec \xi_0 = 0.83 \text{ dB}$$

$$\cos \xi_0 = .8254$$

$$\xi_0 = 34.37°$$

$$\sin \xi_0 = \sin 34.37° = .5645$$

At each elevation beam position in the search pattern use full energy for $34.38° \leqslant |\theta| \leqslant 49°$ and half energy for $0° \leqslant |\theta| \leqslant 34.37°$.

Application of expressions (7.9) and (7.5) for T = 22.2°, θ = 34.37° and N_t = 0° obtains:

Beam Position	E	A_z	α
# 1	0.54°	27.42°	.4604
#8	6.94°	31.96°	.5255
Average			.4929

On the average half energy is used in the search mode for (.4929/.7071 x 100) 69.71% of the beam positions.

The total energy, $\mathcal{E}$, is reduced to:

$$.6971 \frac{\mathcal{E}}{2} = .3486 \mathcal{E}$$

$$.3029 \mathcal{E} = .3029 \mathcal{E}$$

$$.6515 \mathcal{E}$$

The energy saving due to beam scanning management for the search mode is:

(1-.6515) x 100 = 34.85%

A simple computer algorithm can compute θ for every scan angle (expression 7.3) and determine the beam scanning energy management to be used for the search, track and MTI operational modes.

3. *Waveform*

Waveform management is feasible with either a digital or analog signal processor. The number of waveforms used is determined by a cost-effective tradeoff of energy savings versus equipment complexity. Five waveforms are employed in this design. They are obtained by decreasing in binary increments both the pulse length and the time-bandwidth product of a linear fm waveform while the compressed pulse width remains unchanged. The range resolution is unaltered for all operational modes and the receiver bandwidth is the same for all five waveforms. Binary incremental control simplifies the design of the signal processor. In an analog signal processor the various waveforms are obtained by simply switching in-or-out segments of the pulse forming networks.

1) Search

Elevation Beam Positions	Waveform	
	$0° \leqslant \lvert\theta\rvert \leqslant 34.37°$	$34.38° \leqslant \lvert\theta\rvert \leqslant 49$
	μs	μs
# 1 through # 7	55/0.1	110/0.1
# 8	27.5/0.1	55/0.1

2) Track

The following waveform management is obtained by the computer as a function of range and beam scan angle:

Range	Waveform	
nm	$0° \leqslant \lvert\theta\rvert \leqslant 34.37°$	$34.38° \leqslant \lvert\theta\rvert \leqslant 49°$
	μs	μs
200-169	55/0.1	110/0.1
168-142	27.5/0.1	55/0.1
141-120	13.75/0.1	27.5/0.1
119-101	6.88/0.1	13.75/0.1
100-10	6.88/0.1	6.88/0.1

3) MTI

The MTI waveform corresponding to the energy required at 54 nm is:

$0° \leqslant \lvert\theta\rvert \leqslant 34.37°$	$34.38° \leqslant \lvert\theta\rvert \leqslant 49°$
μs	μs
6.88/0.1	13.75/0.1

4. *Power*

1) Search

$\epsilon_{s1} = 78{,}292.5$ joules (Refer to Table 14.1).

with energy management

$\epsilon_s = .6515 \quad \epsilon_{s1} = .6515 \times 78{,}292.5J = 51008J$

where

.6515 = reduction factor

$$\therefore P_{av_s} = \frac{51008J}{2.417s} = 21.1Kw$$

2) Track

Determine maximum average power by assuming 100 aircraft are being tracked at maximum range.

$$\therefore P_{av_T} = \frac{100 \times 110J}{1s} = 11\ Kw$$

5. *Time and Power Budgets*

	Time (Refer to Table 14.1). %	Power %
Search	74	65.7
Track	26	34.3

6. *Duty Cycle*

1) Search

$$D.C_s = \frac{P_{av_s}}{P_p} = \frac{21.1\ Kw}{1000\ Kw} = .021$$

2) Track

$$D.C_T = \frac{P_{av_T}}{P_p} = \frac{11\ Kw}{1000\ Kw} = .011$$

Overall D.C. = .032

3) Short-term Duty Cycle

The short-term and long-term duty cycle transmitter specifications are met by computer controlled duty cycle management. An algorithm is developed to provide an appropriate mixture of search, track and MTI scans per major cycle to meet the duty cycle specifications. A computer major cycle is usually between 0.1 and 0.01 seconds. Another algorithm can be developed to continually monitor the long/short-term duty cycles of the transmitter and not allow them to exceed acceptable design limits.

The following are the short-term duty cycles for the various operational modes:*

	Elevation Beam Position	$0° \leqslant \|\theta\| \leqslant 34.37°$	$34.38° \leqslant \|\theta\| \leqslant 49°$
Search	1	.0212	.0424
	2	.0212	.0424
	3	.0254	.0508
	4	.0309	.0618
	5	.0371	.0741
	6	.0424	.0847
	7	.0483	.0966
	8	.0271	.0542
Track	Range nmi		
	200-169	.0212-.0248	.0424-.0497
	168-142	.0125-.0146	.0250-.0293
	141-120	.0074-.0085	.0147-.0171
	119-101	.0043-.0050	.0086-.0100
	100-10	.0051-.0505	.0051-.0505
MTI		.0086	.0171

*where

$$D.C._{ST} = \frac{T_d}{t_o}$$

$D.C._{ST}$ = short-term duty cycle
T_d = uncompressed pulse length
t_o = interpulse period

14.6 Air Defense ESRS Functional Block Diagram

Figure 14.2 contains the major subsystems of the ESRS. The functional block diagram serves as a road map for describing the operation of the hardware and is a starting point for the development of a full scale schematic diagram.

Table 14-1

Parameter Management Chart

n	E deg	E COV deg	α ms	A deg	β ms	B deg	ξ deg	R nm	T μs	T + 125 μs	$\left(\frac{R_1}{R_n}\right)^4$ dB	$\mathcal{E}$ j	τ μs	DC
1	0.5	0-1.1	707	45.0	-258.5	-15.0	48.8	200	2472.0	2597.0	0	110 (55)	110 (55)	.0424 (.0212)
2	1.5	0.9-2.0	707	45.0	-243.4	-14.1	48.4	200	2472.0	2597.0	0	110 (55)	110 (55)	.0424 (.0212)
3	2.4	1.9-2.9	707	45.0	-228.4	-13.2	48.0	165	2039.4	2164.4	3.3	110 (55)	110 (55)	.0508 (.0254)
4	3.3	2.8-3.8	706	44.9	-213.3	-12.3	47.5	134	1656.2	1781.2	7.0	110 (55)	110 (55)	.0618 (.0309)
5	4.2	3.7-4.8	705	44.8	-198.2	-11.4	47.1	110	1359.6	1484.6	10.4	110 (55)	110 (55)	.0741 (.0371)
6	5.1	4.6-5.7	704	44.8	-183.3	-10.6	46.7	95	1174.2	1299.2	12.9	110 (55)	110 (55)	.0847 (.0424)
7	6.0	5.5-6.6	703	44.7	-168.3	- 9.7	46.3	82	1013.5	1138.5	15.5	110 (55)	110 (55)	.0966 (.0483)
8	6.9	6.4-7.5	702	44.6	-153.4	- 8.8	45.9	72	889.9	1014.9	17.8	55 (27.5)	55 (27.5)	.0542 (.0271)
													MTI	.0171 (.0086)

	Time	Energy	
81 Beam Positions at each Elevation Angle	14076.8 μs x81	825 j x81	
	1.140221 s	66825 j	
3 False Alarms (1 per sec) @ 200 nmi	0.007791	330 j	(3 x 110)
5 MTI Pulses (54nmi) at Beam Positions #1 & #2 = 2x5x81x791.7	0.641277	11137.5 j	(2 x 5 x 81 x 13.75)
(No Time Sharing) Total Search Time	1.789289 s	78292.5 j	Total Search Energy

100 A/C @ 1 Sample per second @ 200 nmi = 100 x 2597 μs = .2597 s/s

Time Remaining for Search = .7403 s/s

$$\text{(With Time Sharing) Total Search Time} = \frac{1.789289}{.7403} = 2.417 \text{ s}$$

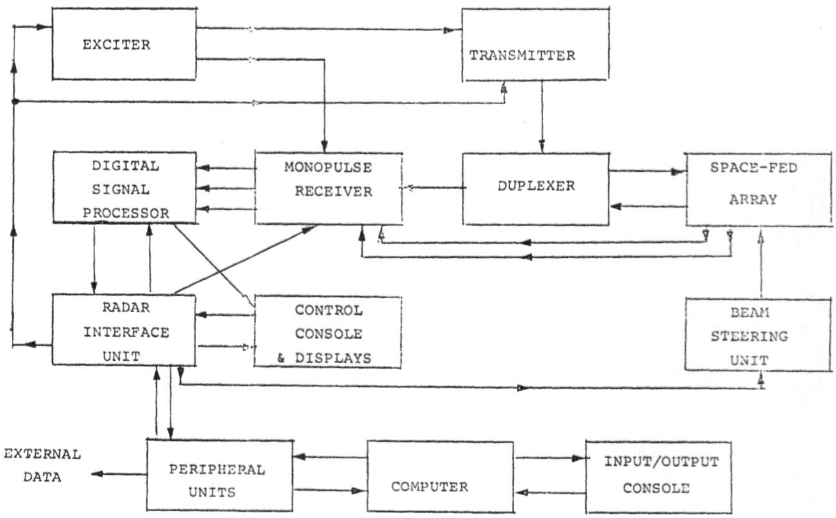

Figure 14-2

Air Defense ESRS Functional Block Diagram

14.7 AIR DEFENSE ESR SYSTEM PARAMETERS

Radar Type	2D ES space fed array 22.2° tilt, 98° conic coverage
Frequency	3 GHz, 10% BW, 10 MHz inst BW
Peak Power	1 Mw
Average Power	32 Kw (DC = .032)
Search Frame Time	2.42s (100 TGT track) 1.79s (no track)
MTI (54nm)	Improvement factor = 50 dB (3 pulse canceller)
(Bottom 2 beam positions)	(5 pulses per beam position)
Track Capacity	50 A/C (1 look/s) & 50 MSL (2 look/s) or 100 A/C (1 look/s)
Coverage	(P_c = 99% P_N = 10^{-7})
($\sigma_{A/C}$ = 1m^2 σ_{MSL} =.05m^2)	(1 false alarm per second)
R_{max}	188 nm (A/C) 82 nm (MSL)
R_{min}	10 nm
Elevation	0° - 7° Search 0°-60° Track
Azimuth	± 45°
Altitude	50,000 feet
Interpulse Period	792-2597 μs variable
Waveforms	
(10 MHz Linear FM)	110 μs, 55 μs, 27.5 μs, 13.75 μs, 6.88 μs
Antenna	Optically fed array 4 horn monopulse feed 50% thinned
Size	7.13 m diameter (hexagonal fit to circular aperture)
Beamwidth	1°
Area Gain	44.0 dB
Losses	4.7 dB
Maximum Sidelobes	28 dB (Taylor taper SL = 35 dB maximum $\overline{n}$ = 5)
No. of Active Elements	5955
No. of Dummy Elements	5955

RMS Sidelobes		42 dB
Null Depth		33 dB (τ = 2.5°)
Polarization		Vertical
Phase Shifters		3 Bit Diode (1.2 dB insertion loss)
RSS Phase Error		20°
Amplitude Error		0.5 dB
Noise Temperature		929° K (2.0 dB uncooled paramp G = 30 dB)
Resolution:	Range	30 m
	Angle	1°
Accuracy:	Range	9 meters (1σ) single hit
	Angle	4 millisines (1σ) single hit
Beam Steering		16 bits input, 4 bit row and column
Beam Switching		0.5 μs
Beam Computation		125 μs
T_x Interpulse Noise		–100 dBm/MHz
Computer Type		X
Speed		1.5 MIPS
Storage		96000 words
Word Length		16 bits
Input/Output Channels		16
Meantime Between Failure		100 HR
Mean Down Time		1 HR
Clock Frequency		80 MHz (rubidium beam standard)
Receiver (Limiting IAGC)		80 dB dynamic range G_1 = 0.5 dB, G_2 = 0.5 dB, α_1 = 5°, α_2 = 10°, α_3 = 2°
Signal Processor Type		Digital
Tracking Algorithm		Type II range and angle

References

[1] L.V. Blake, "Radio Ray (Radar) Range-Height-Angle Charts," *NRL Report #6650*, January 22, 1968; also *Microwave Journal*, October 1968.

[2] Meyer and Mayer, *Radar Detection Theory*, Academic Press, 1973.

[3] DiFranco and Rubin, *Radar Detection*, Prentice Hall, 1964.

[4] L.V. Blake, "Radar/Radio Tropospheric Absorption and Noise Temperature," *NRL Report #7461*, October 30, 1972.

[5] Miller, C.J., "Minimizing the Effects of Phase Quantization Errors in an Electronically Scanned Array," *Proc. 1964 Symp.*, Electronically Scanned Array Techniques and Applications, RADC-TDR-64-225, Vol. 1, pp 17-38.

[6] E.J. Barlow, "Doppler Radar," *Proc. IRE*, pp 340-355, April 1949.

[7] Barton, D.K. & Ward, H.R., *Handbook of Radar Measurement*, Prentice Hall, 1969.

List of Symbols

A_o = antenna area

A_{on} = A sine space coordinate at $A_z = 0°$ and elevation beam positic

A_1 = A sine space coordinate at $A_z = 45°$ and elevation beam position # 1

A_n = A sine space coordinate at $A_z = 45°$ and elevation beam position n

A_z = azimuth angle

A_{zL} = azimuth angle at lower boundary

A_{zU} = azimuth angle at upper boundary

B_{oL1} = B sine space coordinate at lower 3 dB point of beam at $A_z = 0°$ in elevation beam position # 1

B_{on} = B sine space coordinate at $A_z = 0°$ and elevation beam position n

B_1 = B sine space coordinate at $A_z = 45°$ and elevation beam position # 1

B_n = B sine space coordinate at $A_z = 45°$ and elevation beam position n

D = diameter

E = elevation angle

E_1 = center of elevation angle in elevation beam position # 1

E_L = elevation angle at lower boundary

E_{L1} = elevation angle at lower 3 dB point of beam in elevation beam position # 1
E_U = elevation angle at upper boundary
E_n = elevation angle at elevation beam position n
$\mathcal{E}$ = transmitted energy — joules
FAR = false alarm rate
F_1 = preamplifier noise figure
F_2 = mixer noise figure
G = antenna area gain
G_1 = preamplifier gain
G_2 = mixer conversion gain/loss
L = total system losses not included in system noise temperature
L_1 = antenna plus microwave dissipative receive losses
L_2 = losses between preamplifier and mixer
L_{atm} = atmospheric loss
N_{B1} = number of beam positions scanned at elevation beam position # 1
N_{Bn} = number of beam positions scanned at elevation beam position n
N_a = number of active elements
N_t = angle between true north and plane ⊥ to face of array
$(NF)_o$ = operating noise factor
R = range of target — nautical miles
$R_{A/C}$ = aircraft detection range — nautical miles
R_{M1} = maximum instrumented range in elevation beam position # 1
R_{MSL} = missile detection range — nautical miles
R_T = range of tracked target
R_i = initial range where process begins
R_{min} = minimum range — nautical miles
R_{max} = maximum range — nautical miles
R_n = maximum range at n^{th} elevation beam position
S/N = signal-to-noise ratio
$(S/N)_M$ = signal-to-noise ratio at maximum instrumented range

T	= tilt angle of array from vertical
T_A	= antenna sky noise temperature — Kelvin
T_d	= pulse duration — seconds
T_s	= system noise temperature — Kelvin referred to antenna terminals
V_{BS}	= MTI blind speed — meters/second
V_{max}	= maximum velocity — Mach No.
a	= element area
a_{max}	= maximum acceleration — g's
d_e	= effective element spacing in direction where grating lobes first appear
f_o	= transmission frequency — hertz
f_{BS}	= MTI blind speed frequency — hertz
t_o	= interpulse period
Δ	= amount of stagger — fraction
Δ_{ATD}	= decrease in atmospheric loss between R_i and R_n
a_{on}	= a sine space coordinate at A_z = 0° and elevation beam position n
αm	= maximum angular acceleration
a_1	= a sine space coordinate at A_z = 45° and elevation beam position # 1
α_m	= maximum angular acceleration
a_n	= a sine space coordinate at A_z = 45° and elevation beam position n
β_{on}	= β sine space coordinate at A_z = 0° and elevation beam position n
β_1	= β sine space coordinate at A_z = 45° and elevation beam position # 1
β_n	= β sine space coordinate at A_z = 45° and elevation beam position n
δ	= amount of stagger — seconds
ϵ_{MTI}	= energy reduction at MTI range
ϵ_{Rn}	= energy reduction at n^{th} elevation beam position
ϵ_{R_T}	= energy reduction of tracked target

θ_o = 3 dB beamwidth

θ_L = maximum scan angle from array normal at lower boundary

θ_U = maximum scan angle from array normal at upper boundary

λ = wavelength of transmission — centimeters

ν = 1.06711 (.28 dB) conversion factor

σ = target cross section — meters^2

ξ_1 = scan angle from array normal at A_z = 45° and elevation beam position # 1

ξ_{max} = maximum scan angle from array normal

ξ_n = scan angle from array normal at A_z = 45° and elevation beam position n

τ = precomparator phase unbalance between two ports

τ_e = equivalent pulse length — seconds

Index

www.ingramcontent.com/pod-product-compliance
Lightning Source LLC
Chambersburg PA
CBHW051255200926
51612CB00001B/49
* 9 7 8 0 8 9 0 0 6 0 2 3 0 *